Norman Potter

What is a designer
: things . places . messages

Hyphen Press

© Norman Potter, 1969, 1980, 1989

1969 first published by Studio Vista (London)
and Van Nostrand Reinhold Company (New York)

1980 revised and extended edition published by
Hyphen Press, Reading

1989 this revised third edition published by
Hyphen Press, London

new material set by Merrion Press, London

made and printed in Great Britain by
WBC Print Ltd, Bristol

ISBN   0 907259 03 0

To Mark, † 1969, and those with him at Bristol, Hornsey, Guildford

Sequence of parts

Architecture is organization. *You are an organizer, not a drawing-board stylist.*
LE CORBUSIER

Woe to the man whose heart has not learned while young to hope, to love – and to put its trust in life.
JOSEPH CONRAD

The disciplinary barriers are impenetrable. If these barriers in education were to vanish, the architect as benevolent dictator would vanish too. Instead students could arm themselves with useful tools and knowledge with which they could assist a community.
TOM WOOLLEY

Sources for these and other references are given in part 27

Introduction

This is book for students who design artefacts of the kind studied in design and architectural schools. Aside from the question implicit in its title, it asks what skills and aptitudes may be appropriate to the practice of design. A general discussion of such matters, including detailed recommendations for further reading, is followed by a directly practical reference section, confined to essentials, on various aspects of design technique (distinguished by a change of page colour). The book concludes with a survey of information sources, listed useful addresses, and some advice for absolute beginners. The absence of pictures is deliberate and considered. The reasons will become apparent to any thoughtful reader, as the argument of the book develops, but there is a terse summary of this reasoning in part 23 ('Questioning design'; see pages 197-8 particularly).

By the word 'student' I mean to include those of all ages coming freshly to their subject. At a certain level of awareness all creative workers gain in humility as their knowledge develops, and will wish frequently to return to their origins for refreshment of the spirit: to ask yet again who they are, what they could or should be doing, and why. I would like to think that this book will be a congenial companion to any engaged by such a quest, as much as to those who, starting their journey, turn to it for guidance.

What is a designer now enters its third decade in this revised third edition. The addresses and similar references have been carefully checked through and updated where necessary. The reading list could well have been vastly extended – but was not. Almost any such list is already too long, either by pre-empting discovery, or discouraging those who may be excellent designers, but less than ardent in their reading habits. The suggestions offered are still, I believe, sound and stimulating as foundations to any house of theory worth living in, and just now it seems that secure foundations are rather more necessary that plausibility of superstructure. There is also, of course, a designed provision for growth and change in the life of the book. New readers will discover this both in the suggested 'categories' for a designer's reading and in the notespace provided for their own use. The full occupation of this space will not only be defensibly territorial (to the interests of its owner) but will certainly help the active life of the book as a working tool.

A mindful question is whether this third edition has been rewritten to take sympathetic account of a newly-emerged entrepreneurial and allegedly pluralist design culture, closely associated (in Britain) with the successful ripening of Thatcherite social policies. The answer is no: the book stays as it is. Its message was hardly fashionable to begin with. Whether or not a society gets the design it deserves, the 'designer culture' of the 1980s reflects its social background with uncanny (and unnerving) precision. Possibly a more tough-minded opposition might have helped (or might yet). It is true that the question 'what' is the moment of decision for a designer, distinguishing those who practise design, with every consequence entailed, from those who use, teach or criticize both the practice and its tangible outcomes. It is creatively desirable, this book suggests, that moments of decision become also moments of truth – leverages in that regard; whether personal or social. The 'why' and the 'how' in design work are separated only at severely disabling cost to its social meaning and intrinsic worth.

It is well enough understood that design is a socially negotiated discipline, and there are telling respects in which design questions are political questions. No book about design is politically value-free, whatever its apparent claim to objectivity. I would like to make it plain, therefore, that this book has always belonged resolutely with the standpoint of the libertarian left, and still does so. It is a textbook of design, not a political tract; but that is its standpoint. It also adopts and explores – even more unfashionably – a generally modernist position in that context, holding that 'like the poor (it seems), what is modern is always with us; and open to transformation'. Constructive forces in the world should not be confused with those of reaction, but there is nothing very new in such conflicts; nor in the difficulty of their interpretation.

There are certain very general facts of twentieth-century life within which any constructive activity must be seen as contributory. Untouched by them, any such activity is perhaps better suited to garden gnomes than to human beings. These facts include a global view of the human situation, which can hardly exclude the decimation of the rain forests any more than the appalling disparity of living standards, and life expectation, in the different parts of our world; not to speak of the nerve-centres in this century's self-awareness, which have been the revelations of the concentration camps and the terrible lurking ever-present destructive threat of the Bomb—now so inconceivably destructive even in one warhead as to defy imagining. Yet if human solidarity and humane imagining mean anything at all, it is upon such awareness that every act of construction, however small, must in some sense draw, seek nourishment, become predicated—or withdraw into triviality and the blandishments of excess. A certain order of sensibility, if not of commitment, belongs to such perceptions; and a different one, to their suppression, or neutering. To expose and to clarify, not to embellish, was at once the joy and the seriousness of the modern movement, and it remains vital to its task and heritage.

To some of these questions, and to the underlying problems of industrialism as such, E. F. Schumacher has brought a notable diagnostic acumen as much as simplicity of heart, putting in very simple language what must surely be plain: 'There are four main characteristics of modern industrial society which, in the light of the Gospels, must be accounted four great and grievous evils:
1. Its vastly complicated nature.
2. Its continuous stimulation of, and reliance on, the deadly sins of greed, envy, and avarice.
3. Its destruction of the content and dignity of most forms of work.
4. Its authoritarian character, owing to organization in excessively large units.
All these evils are, I think, exacerbated by the fact that the bulk of industry is carried on for the purpose of private pecuniary gain . . . (but) the worst exploitation practised today is "cultural exploitation", namely, the exploitation by unscrupulous moneymakers of the deep longing for culture on the part of the less privileged and undereducated groups in our society.'

Schumacher also had constructive offerings, not necessarily to the taste or within the aptitude of many designers, but since in this book I am arguing that 'design is a field of concern, response, and enquiry as often as decision and consequence'—big words!—it may be encouraging to remember that we still have visible role-models within a radical Christian attitude. Trendies take note.

It remains to ask, in an introductory spirit, what is special about this book, and justifies a third edition. First, it is original: that is to say written at first hand from my own experience as a designer, maker, and teacher; rather than from reading other books. It is useful, I think – useful evidentially to students – to have felt the modern movement in your bones, to have lived it. Such 'being original' is not exclusively useful, however. Creative critical scholarship – to which I have little claim – has something different to offer, and complementary: the special benefit of breadth, distance, and relative impartiality of view. This brings me to a second foundation asset of this book. Like – I suggest – the root impulse and need of truly modern design, it is not self-contained; it is contributory. There are other books by scholars, and many books of pictures. The first thing to learn about the deep structure of modern design is that it is *relation-seeking* and pleasurably so; but *that*, as the saying is, we shall come to. Good work was probably always difficult, and recently never more so; but I hope this book will encourage those who hold to it.

Norman Potter/1989

1 What *is* a designer?

Design:
v. to mark out; to plan, purpose, intend . . .
n. a plan conceived in the mind, of something to be done . . .
n. adaptation of means to end . . .
The shorter Oxford English dictionary

Every human being is a designer. Many also earn their living by
design – in every field that warrants pause, and careful consideration,
between the conceiving of an action and a fashioning of the means
to carry it out, and an estimation of its effects.

In fact this book is concerned mainly – not wholly – with a minority
profession: of designers whose work helps to give form and order
to the amenities of life, whether in the context of manufacture, or of
place and occasion. The very clumsiness of this definition underlines
the difficulty of using one word to denote a wide range of quite
disparate experiences – both in the outcome of design decisions, and
in the activity of designing. The dictionary reference above is
selective; in practice the word is also applied to the *product* of 'a plan
conceived in the mind', not only as a set of drawings or instructions,
but as the ultimate outcome from manufacture.

This is confusing. The difficulty becomes acute if the word 'design'
is used without reference to any specific context – used, for instance,
as a blanket term to cover every situation in which adaptation of
means to ends is preceded by an abstract of intent – though designing
is thus usefully distinguished from 'making' or from spontaneous
activity. Beyond this point, the word must refer to recognizable
products and opportunities, or become hopelessly abstract.

The design work to be discussed is now usually studied (in Britain)
within the art and design faculty of a polytechnic, in a school of
architecture, sometimes within a university, and – not least – in some
of the smaller art and design colleges that may provide vocational
courses. The Open University provides study courses, necessarily at
a generalized level and without the benefit of studios and workshops.

And of course, with or without the aid of such studies and evening classes, it is perfectly possible to study design simply by doing it. It should not be necessary to say that architects are designers (even if the matter is, occasionally, in doubt). Taking that old stand-by the 'broad view', it is convenient to group the work into three simple categories, though the distinctions are in no way absolute, nor are they always so described: product design (things), environmental design (places) and communication design (messages). Such categories blur some further necessary distinctions (as between, for instance, the design of industrial equipment and that of retail products in a domestic market) but can form a useful departure.

In the field of product design, the professional extremes might be said to range from studio pottery and textile design at one end of the spectrum, to engineering design and computer programming at the other. This is a very broad spectrum, and clearly there are serious differences at the extremes. In the communication field, a similar spectrum might range from, say, freehand book illustration, to the very exact disciplines of cartography or the design of instrumentation for aircraft.

Obviously, the more aesthetic and sensory latitude available within a particular range of design opportunities, the closer they resemble those offered by the practice of 'fine-art'. The less latitude, the closer design becomes to the sciences, and to fields in which the scope of aesthetic 'choice' is truly marginal. The design of a traffic light system has an aesthetic component, but it would need a very special definition of aesthetics to embrace the many determining factors that must finally settle the design outcome.

The situation for architects is usually held to be more straight-forward; historically, their position has developed a fairly clear set of responsibilities. However, the complex changes in building types, and in industrialized building possibilities, have combined with other factors thoroughly to upset this stable picture. Indeed, the architect's work has been so undermined by that of specialists in surrounding territory (engineers, planners, sociologists, interior designers, etc.) that the profession is no longer so easy to identify. It is still reasonable to see an architect as a designer with a specialized technical and functional competence, and again a spectrum is discernible, ranging from very open and ephemeral design situations, to those as critical as the design of an operating theatre.

It is necessary to start somewhere, and this book takes a middle-zone standpoint. In most art schools this will include furniture, interior design, exhibition design, packaging, some wide areas of graphic and industrial (product) design, and some of the fringe territory leading into architecture. Students must make the necessary allowances to accommodate their own subject of study. This is chiefly necessary in part 2 ('Is a designer an artist?') and in some of the notes on procedure (parts 12-17) – or the studio potter will certainly feel that everything in this book is unduly complicated, whereas an architect might feel that there is undue simplification. All designers, however specialized, should know roughly what their colleagues do – and why; not only to fertilize their own thinking, but also to make group practice effective, and for other reasons that will appear.

There are many roles for designers even within a given sector of professional work. A functional classification might be: impresarios, culture diffusers, culture generators, assistants, and parasites. Impresarios: those who get work, organize others to do it, and present the outcome. Culture diffusers: those who do competent work effectively over a broad field, usually from a stable background of dispersed interests. Culture generators: obsessive characters who work in back rooms and produce ideas, often more useful to other designers than the public. Assistants: often beginners, but also a large group concerned with administration or draughtsmanship. Parasites: those who skim off the surface of other people's work and make a good living by it. The first four groups are interdependent, necessary to each other. It should be added that any designer might shift from one role to another in the course of his working life, or even within the development of a single commission, though temperament and ability encourage a more permanent separation of functions in a large design office. Thus no value-judgement is implied here; except upon parasites who are only too numerous.

In small offices – or of course for independent free-lance workers – there will be little stratification; 'the office' may tend to move in one direction or another, but the work within it will be less predictable for any one member – excluding, perhaps, secretarial or administrative assistants and often temporary draughtsmen. A 'consultant' is often a lone wolf who deals in matters of high expertise or (paradoxically) of very broad generality. Designers will be found in every quarter; sometimes working independently, sometimes for government or local authority offices, or attached to large manufacturers, to retail agencies, to public corporations, and elsewhere in places too numerous to mention. Artisan designers will have their own workshop and perhaps their own retail outlet. There are a few design offices that will design anything from a fountain pen to an airport, and will therefore employ specialists from every field (including architects) – a rational development and a welcome one, but implying some genius for large scale organization which, in turn, may tend to level out the standard of work produced. (As numbers increase, it becomes a problem to keep work flowing through at a productive pace, yet have enough – not too much – to allow everyone a fair living.) Students usually need a few years' office practice before setting up by themselves; often this happens in small groups of three to six designers who will share office and administrative expenses.

Most designers are educated in a formal way by three-to-seven years in a design school (or school of architecture) leading to appropriate qualifications. Some have had unorthodox beginnings – by dropping in the deep end and learning to swim – but self-training may need sympathetic patrons, is apt to be patchy according to the opportunities that occur, and needs a special pertinacity. Apprenticeship rarely means more than training as a draughtsman. A few factories or retail firms may encourage employees who show design aptitude. Evening classes and correspondence courses are mostly directed at cultural-appreciation or do-it-yourself horizons, but intending full-time students can build up a portfolio of work by this means.

A note of warning: the word 'design' appears freely as noun and verb, and where words like 'formal', 'realization', 'consciousness' are used without qualification, readers should examine the context and think for themselves. I have used the word 'student' suggestively; trying it for size.

There is a perfectly good sense in which a creative worker remains perpetually a humble student of his subject. This is not to be confused with the timidity of the 'permanent student' whose name haunts the lists of application for grants, research funds, and finally, minor teaching appointments. These must again be distinguished from the serious student of scholarly bent who 'reads' the subject and may make a distinctive contribution to theory or criticism. By the word 'student', therefore, I mean those who still question what they are doing, and ask why.

There is no word by the use of which sex-discrimination can be avoided. Readers must accept that when 'him' or 'man' is used, these words embrace both sexes (unless the text does draw a distinction). Women should not be deterred from course-work that includes the use of machinery and unfamiliar work with hand-tools. Invariably such skills are gained rapidly and practised with enthusiasm.

This, then, is the apparent situation of the designer and where this book begins. Returning to the statement that every human is a designer, and using it as a springboard: we do well to remember that designers *are* ordinary human beings, as prone as others (given half a chance) to every human weakness, including an exaggerated idea of their own consequence. Consider the following questions: Should a designer design for a factory in which he could never imagine working as an operative? Is design social-realist art? Is it handy to be in a state of moral grace when designing a knife and fork? Does design work justify its claims to social usefulness, or is it a privileged form of self-expression? Is a profession a genteel self-protection society with some necessary illusions? Should a designer be a conformist or an agent of change?

Those who feel that such questions are diversionary and a waste of time, should perhaps put this book down; others read on, but not for easy answers.

2 Is a designer an artist?

Before discussing this question, which involves describing a designer's work in some detail, it is necessary to look at the context in which it is usually asked. In Britain, it is certainly necessary to remember two things. First, the extraordinary cultural insularity of the last fifty years that permitted the early achievement of the Arts & Crafts movement to be built into foundations of growth on the continent, while continuing a placid, homespun and largely arrested development in its place of origin. To illustrate this, it is only necessary to consider the lively interaction of media, disciplines and controversy in the *de Stijl* movement, and to take a sample of the English situation at the time, or to recall the strange scene in 1968 when so many students and teachers were to be seen confronting their first substantial awareness of the Bauhaus. The book *Circle,* published in the late 1930s, marked a coming-together which, in British terms, was promising merely because unusual. An informal pointer to an absence of spirit can be seen from a comparison of the early edition of Herbert Read's *Art and industry,* with layout by Herbert Bayer, and the subsequent 'tasteful' editions in which the spirit of the original – not in itself anything very remarkable – is absorbed back into the British literary traditions of book publishing. It was indeed in literature that things were more lively. The war destroyed even these tentative growths in a way that was more profound than is generally realized. The Festival of Britain (1951) and its major Exhibition became a rallying point for a new design consciousness. Although in some ways a joyous occasion and no mean achievement for all concerned, it proved to be a very odd mix of English empiricism with a belated tribute to the 'international style' in architecture. There was however a distinctively fresh sense of place and occasion on London's South Bank, to set against – or more hopefully, to assimilate – the threat of foreign invasion. The implicit fear that without a struggle everything might come to look alike, was indeed soon to be justified. In matters of educational debate, however, the situation was far from robust. I can testify to the difficulty of interesting anyone in a confrontation with Tomás Maldonado, of the Hochschule für Gestaltung (Ulm), at the Royal College of Art; and this was 1961.

The second difficulty is more widespread. It is the well-known but uneasy juxtaposition of 'fine-art' studies with 'design' subjects within a common faculty, excluding (normally) architecture. It would be out of place here to examine the history of this problematic and to some extent (now) arbitrary grouping of studies. It is enough to point out that the situation could be more realistically appraised if painting and sculpture were studied alongside music, dance, poetry, film and other activities that interpret, primarily, the psychological and sensuous and spiritual understanding of man. It would then be easier to distinguish those activities which must first satisfy his physical and accessory needs under conditions of complex social constraint (as in building design), or which may have a much humbler role in serving and pleasing man. It is true that, in the last analysis, every human artefact – whether painting, poem, chair, or rubbish bin – evokes and invokes the inescapable totality of a culture, and the hidden assumptions which condition cultural priorities. (In a basic sense, and given the conditions for warmth, food and shelter, the rest is a choice and speaks to us of priorities which need constant revaluation.) For the purpose of the remarks which follow, it is certainly necessary to say that if the words 'fine-art' and 'design' simply refer to a duality as experienced in art schools, it is difficult to set up satisfactory distinctions on that basis alone.

For the discussion that follows, the situation is seen from the standpoint of a designer.

Here is a sober but accurate description of professionalism by Professor Misha Black: ' ... the offering to the public of a specialized skill, depending largely upon judgement, in which both the experience and established knowledge are of equal weight, while the person possessing the skill is bound both by an ethical code and may be accountable at law for a proper degree of skill in exercising this judgement ...'

Not, obviously, a full description, and perhaps a somewhat negative one, but making the fact plain that a designer works through and for other people, and is concerned primarily with their problems rather than his own. In this respect he might be seen as a medical man, with the responsibility a doctor has for accurate diagnosis (problem analysis) and for a relevant prescription (design proposals), though the comparison should not be taken too far. It must be clearly realized that designers work and communicate indirectly, and their creative work finally takes the form of instructions to contractors, manufacturers and other executants. The exception is the designer-craftsman or artisan, whose situation is discussed in part 7. The instructions may include written specifications, reports, and other documents, detailed working drawings, presentation drawings for clients, scale models and sometimes prototypes in full size. Since this is as far as a designer goes in direct production (strictly what he makes are visual analogues), it is necessary that the instructions are very clear, complete, and in other ways acceptable to those who must work from them. Much is said about this requirement in parts 12 to 17 of this book.

The designer usually has the further responsibility of supervising the work, but there is no obvious equivalent for the feedback through eye-and-hand so familiar to the painter or sculptor, whereby the original idea is constantly developed, enriched, or diverted by the actual experience of the materials and the making-process. The artisan is an exception. For most designers the point of no return (commitment to final drawings) is indeed final, unless everything is upset by site contingencies. So-called 'feedback' does of course operate at the design stage, mainly through people, circumstance, and the continuous absorption of new information into the design brief, which will alter its definition. The outcome will still change radically from first ideas thrown up by superficial acquaintance with the design problem, but the changes will not always be of the designer's own choosing: their nature may be objectively determined by factors quite outside his control. Such factors might be something to do with costs, the availability of materials or techniques, a change in the client's requirements, or simply the discovery of factors that were hidden from sight in the early stages of the job.

Hence, in summary, the designer provides instructions (having exhaustively established and agreed the best course of action), and the work necessarily involves many different people whose interests (often in conflict) he must seek to reconcile. With some such people he may have (legally) contractual relationships.

Thus many specific responsibilities may arise – to clients, contractors, to the public who use the end-product, to numerous specialists or colleagues who may be involved if the undertaking is a large one (which implies team-work and frequently shared decisions). If it is a building, it mustn't fall down; if it is a chair, it mustn't be thirty inches high, have an innate tendency to collapse under load, it mustn't employ joints that can't be made except by special machinery (unless this can be found economical) and it mustn't cost so much as to be unmarketable. The designer cannot exercise personal insights until every apparently conflicting factor in his brief has been reconciled to best advantage: until, in short, he knows exactly what he is up against and which constraints can be made to play in his favour.

For such reasons, the designer is highly 'problem' conscious; a large part of his work may consist in problem analysis, though rarely of the complex order familiar in the sciences. To an ability for sorting, ordering, and relating information he must bring qualities of judgement and discrimination as much as a lively imagination. There is a diffuse sense in which the most seemingly 'objective' procedures in problem analysis are in practice discretionary, embedded as they are in the whole matrix of professional judgement in which relevant decisions are conceived. In some fields (such as textile design) there is far greater latitude than in others. In most design work the ultimate decisions affect, in a vital degree, appearance; but the look of the job, however lovingly considered, will emerge from, and in some sense express, the functional and circumstantial background. There are of course cases in which a communication requirement will be superimposed overridingly upon other factors, like structural logic; that would simply be a special (perhaps sophisticated) view of function.

Drawings can never be an end for a designer (excepting an illustrator); they are a means to the end of manufacture, and their expressive content is strictly limited to the purposes of relevant communication. This obvious distinction from fine-art drawing can easily be overlooked in a design school where the design projects are theoretical, and drawings become the only outcome, acquiring the false dignity of an end-product in the process. This does not imply that drawings can be loveless, slovenly, or inadequate in any way, but that their nature is strictly purposeful. It may indeed be necessary to the designer to make loving, scrupulous and over-adequate drawings for his own self-satisfaction and to preserve his own standards. Only in this sense are design drawings 'self-expression'.

At every stage of design there will be discussion, questions and argument; the final design will have to be demonstrated and if necessary defended to the client, who will not understand what the final result will look like, but will naturally tend to assume that he knows more about his own problems than does the designer – despite having called him in to solve them. A design proposal intermingles with the world of considerations familiar to the client; communication media must be carefully chosen – verbal reports and other documents may accompany drawing and models. Designers use words constantly and in direct relation to their work; in forming and discussing ideas, assessing situations, annotating drawings, writing specifications and letters, and in report writing. This aspect of design work is frequently underestimated: an ability to use words clearly, pointedly, and persuasively is at all times relevant to design work.

It is now possible to ask, what kind of person might be happy and personally fulfilled in taking up design? It will be seen that a designer must be capable of more detachment than may be necessary to a fine-artist. He must be able to weigh up a problem, or an opportunity, in a dispassionate way, on its terms (as well as his own), and to select, arrange, and dispose his decisions accordingly. He must be able to thrive on constraint and to turn every opportunity to good account. He must like and understand people and be able to treat with them; he must be able to accept fairly complex situations in which he may well be working as a member of a team. He must be reasonably articulate. He must be practical and prepared for extensive responsibilities to other people. Finally he must be prepared to spend at least half his time working with graphic media, since most design work appears in drawings of one sort or another when decisions have been finalized.

These remarks may suggest an uncomfortably glum idea of human perfectibility. In practice, of course a designer's life is as muddled, informal and accident-prone, as most people's lives manage to be; not only behind the scenes, but sometimes in front of them. Every profession has roughly defined public responsibilities, which are met as closely as possible by accepted codes of practice. Again, the fact that design work is ten per cent inspiration and ninety per cent fairly hard work – not an unusual prospect – does need some well-organized procedures to keep the brief clearly in view, and the available energies best occupied.

Some of these procedures will be familiar to painters and sculptors, and certainly to film-makers; but for them the work will have a more inward character in its origins. Thus a painter's first responsibility is to the truth of his own vision, even though that vision may (or maybe always does) change as his work proceeds. He may be involved with contractual responsibilities, but not to the same extent as is a designer, whose decisions will be crucially affected by them. The designer works with and for other people: ultimately this may be true of the fine-artist, but in the actual working procedure a designer's formative decisions have a different order of freedom. The fine-artist is less dependent on discussion, agreement, letters, visits: the apparatus of communication that brings definition to a design problem, and relevance to its solution. A fine-artist usually works directly with his materials, or with a very close visual analogue to the final work. As we have seen, the designer has a long way to go before firm proposals can emerge – and even then a model may be the nearest thing to a tangible embodiment of his ideas.

In the case of film, television, and theatre, which might be described as a realm of public art, quite complex design procedures are involved. In the main, however, the real connection between fine-artists and designers springs from the benefit of a shared visual sensibility; not from a relevant or direct transference of skills, language, or formative insight, from one field to the other. Students are warned that this is an opinion: recalling the breadth of the design 'spectrum', they will see that this is a difficult matter to unravel. So many factors impinge on the visual appearance of a design outcome that a designer's hand would seem to be guided by a wholly different 'requiredness' (a term borrowed from Gestalt psychology) from that which informs a painting or a work of sculpture. Yet there are component experiences with something in common. Equally valid transferences may occur from the 'feel' of related work in other fields (for example, philosophy, music, or mathematics) and should be encouraged to do so. Similarly, a creative sensibility may derive from unlikely sources that cannot be looked for in any one field alone.

It is only necessary to hammer home the obvious because fine-art and design (excepting architecture) are often taught as closely interrelated subjects, and students are asked to choose between them. The isolation of architecture, which has always been the home base of design theory, is hard to explain and justify. Perhaps the theory is affected by it. The term 'fine-art' is unpleasantly genteel, but will be met with in the careers prospectuses and in the art schools, normally to comprise painting, sculpture, printmaking, and photography, and to distinguish these studies from 'applied art', in the various fields of design discussed here. The view that there is a parallel situation in the sciences, as between pure science and applied technology, is a questionable one: equally untrustworthy is the supposition that painting, sculpture, industrial design, architecture, derive in some sense from the common fountain-head of 'art'. To suggest this seriously requires a view of art (and a set of definitions) quite outside the scope of the present discussion: it is partly a semantic problem, pointing to the inadequacy of ordinary descriptive language. Without distorting common usage, it might be said that designers are content to bring a certain artistry to their work, and to recognize that there is much in common between the few masters in any field – fine-art, design, science, medicine, philosophy – more, perhaps, than unites the very disparate standards that co-exist in any one profession.

3   Design education: principles

'Well building hath three conditions: commoditie, firmnes, and
delight' (Vitruvius/Sir Henry Wotton)
'Love, work and knowledge are the well-springs of our life.
They should also govern it' (Wilhelm Reich)

A design capability proceeds from a fusion of skills, knowledge,
understanding, and imagination; consolidated by experience. These
are heavy words, and they refer to the foundations. We accept a
certain minimal competence as the basis of professional self-respect,
and as some guarantee of a designer's usefulness to other people.
Within limits such a competence is definable, and will begin to form
outlines within a formally structured teaching/learning situation.
It is too much to say outright that design ability can be 'taught'. As
with any other creative activity, it is a way of doing things that can
only be grown into, perhaps – but not necessarily – in the context of
a formal design education.

This view is readily conceded for something as immaterial as
'imagination', but it is commonly held that skills and knowledge must
not only be taught, but rigorously examined: if only to protect an
unsuspecting society against social or technical malpractice.
Defensible as this may be, it is not an assumption that should go
unquestioned, nor deflect attention from the weaknesses of received
professional standards. The damage caused by knowledge used
without understanding is merely difficult to measure: it is not less
real for that. A skill may be irrelevant to the nature of a problem, or –
in dealing with people – may be grossly uninstructed in a necessary
tact and discernment. Knowledge may be thinly experienced as a
rag-bag of conventional responses helped along by access to someone
else's published working details. Plainly, skill and knowledge cannot
be weighed out by the pound, and separated from qualitative
perceptions, for any but the simplest mechanical problems – and even
there it is questionable. Even 'judgement', that wise old word, becomes
ponderously inhuman unless fertilized by some order of creative
spontaneity.

Architecture is only one profession that offers uncomfortable testimony in these respects. It is nice to know, for instance, that a building is unlikely to leak or fall down (in fact someone else has probably done the calculations) and in a simple way this must be counted a social gain. On the other hand there are very many thousands of architects all of whom have passed their examinations, and can we decently say that more than a substantial minority produce buildings of affective quality? Of course their work is sometimes exceedingly difficult and subject to every conceivable restriction, but does their education help them to feel desperately a gap between promise and fulfilment, and thus to find every conceivable way of bridging it?

Let us answer (to be charitable) that the best schools know this problem backwards and do their best to resolve it. A 'profession' can still become a self-protection society with a very short term view of the priorities for professional competence. In the long term, we have no yardstick for the spirit of man and the nature of deprivations in his environment – not only the wilderness he may see and accept all around him, but the very nature of his interaction with a wholeness of experience, of which a built environment is but a part.

These are large thoughts, and will surely bring a smile to the face of any over-worked architectural assistant, aware as he is of the drudge component that occupies so much of his conscientious labours. Yet to evaluate formal design education it is necessary to ask some awkward questions. Not only to disturb our unthinking acceptance of social norms, but to bring some very practical matters sharply into focus, and others to dismiss as marginally relevant. For instance, our assistant would confirm that much of his work is elaborately interwoven with building contingencies and the structure of 'consents' which sometimes keeps him awake at night; the personality of the local district surveyor may occasionally seem the most omnipresent factor in the whole of the job. It is impossible to explain this properly to a student in a school; you simply need the experience of design practice to see how it happens, and just what you do to keep the job moving and your first intentions reasonably intact, or, as is often the case, subtly changing as new possibilities reveal themselves. The way in which accidents of site contingency suddenly appear as benefits, so the designer wonders why he hadn't thought of that in the first place – to explain *that* as part of the 'design process' is really not at all easy.

It is also desirable (to put it mildly) to see the educational problem in terms of the future as much as the present – and, in rather a different way, in terms of the past. It would be a mistake to pre-suppose a static social situation, pleasantly unified in the untroubled pursuit of affluence, disturbed at most by some new concession to the good life announced in the weekly journals, and the designer providing his expensive austerities where they can be afforded; namely, in the places of high financial decision-making. Such banalities may go sour before the last Barcelona chair is glued into position to impress – who? There were signs in 1968, if not now, that young people were certainly less impressed than some of their parents. Yet a measurable standard of living is not to be despised because not all of us have it or want it, when millions of people desire above all some alleviation of their physical poverty. Nor can we play God and start back from square one with a wholesale redistribution of resources and an imposed system of moral absolutes to keep everything tidy. In short, the designer, like other honest citizens, will need access to faith and vision as much as to a keen analytical intelligence – to engage with life effectively, and to make something good through work. Aside from a technical training, what can a design school do to help?

This is not easy; a simplistic approach helps nobody. Hence the paragraphs that follow.

A first requirement for students is knowledge of how they can best help *themselves:* in this respect it is useful to understand the limits and benefits of an academic situation. To begin with, students should realize that both education and design practice are too often handicapped by identity-fixations. The words by which people describe themselves – architect, graphic designer, interior designer, etc. – become curiously more important than the work they actually do. In one respect this is fair, because under modern conditions it may be very difficult to find one word to identify their work, but such words tend to build up irrelevant overtones of meaning which are more useful as a comfort to personal security than as a basis for co-operative enterprise. Such confusions interpenetrate with status values and the other intricate strands of our social life, so it is hardly surprising that education is affected by them. Thus it comes about that design education is often still irrationally divided up into specializations with a doubtful relation to the work students may finally do, and with even less plausible reference to the situation *as it could be* in ten years' time.

This is not in itself the case for a generalist education – the silliness of outdated categories is a negative consideration – nor can matters be improved by denying that specific subjects relevantly exist, and that they are worth studying in depth. In theory, of course, a three-year study programme could investigate the Universe from the confines of a single problem. It is also a fallacy of bad technical training – as implied earlier – to suppose that skills and knowledge can be picked up *in vacuo* or in neat packages as in a supermarket; such knowledge is sometimes used with a contempt and a restlessness that betrays the additive nature of its learning. Thus a restless and dissatisfied student who seeks his freedom in the extent, rather than the depth, of his explorations, may be dancing to the tune of the masters he is trying to escape from (whether such masters are real or imaginary). Skills ought to keep pace with understanding, and thus emerge organically; and the spread of lateral thinking must be complemented by vertical thinking (and dreaming) of a quite different kind, if understanding is to grow from strong roots. It is also a mistake to think that good design (or creative thinking of any kind) has much to do with a perfectly fitted-out environment. Much design work can be approached with freshness and insight with very little in the way of equipment and materials: the assumption to the contrary (as an accepted starting point for design) merely reflects a social climate in which the least meaning emerges for a gross expenditure of effort and apparatus.

It is a sad but interesting observation that the most rigidly blinkered courses of 'training' are often the least *technically* competent, rather in the way that some committed amateurs do better work than tradesmen who have lost a responsible relation to their trades. (*Amateur* – one who loves . . .) Even if it were not so, an education should always be why-responsive (and thus, question itself), whereas of its nature, a training asks *how,* and provides tidy answers. In either case, however, the best theoretical framework for studies will be alive or dead rigid, according to the spirit in which it is interpreted, both by students and staff, which in turn is a function of confidence within the academic community.

Although this discussion is tending to assume that the schools and colleges are where it all happens, Ivan Illich reminds us that it is easy to confuse schooling with education, when in fact education is a natural life-process and some schools may actually be anti-educative. A certain realism of outlook is required, if the schools are to be seen for what they are. A design school may be in some sense preparatory for life to follow, but a student's time there is precious and irrecoverable life in the present. It is also, however, a safe berth in harbour. In a somewhat different way, the staff of a school seek their own expression in the achievement of the school; it is their own chosen and continuing way of life. On the whole, the work is agreeable, relatively well paid, and socially well regarded. It is work that allows some teachers to defer indefinitely a close look at their own inadequacies. For many reasons – not all so uncharitable – an educational *connivance* becomes possible between students and staff which may give a wholly false emphasis to the importance of structured education. There is a good sense in which any designer worth the name will be a student for the whole of his working life. This is not only a function of creative resource, but also of the conditions of rapid technological change which a designer must meet in his work. In such a perspective, the few years in a design school are not unimportant, but should be carefully guarded against inflated claims for 'completing' the education of a designer. At best, three – or even five – years in design school can attend (thoroughly) to a few simple priorities in a designer's personal education. Even at a mundane level, much of what is called 'operational know-how' is necessarily picked up in the rub of professional practice.

What, then, is the special usefulness of a formal design education? Numbers of staff and students are gathered together in one place with a common purpose and common facilities. Students learn a great deal from each other; not only from books or from their tutors' guidance. These facts point to the benefits of academic life: a gradually widening area of agreement (the norms against which individuality becomes meaningful), the experience of sharing, co-operating, and resolving conflicts: in a word, the chance of *participation* in all the stress and stimulus of a particular community with shared aims. This side of 'further education' is no less formative than the acquisition of skills and knowledge, much of which (not all) is equally available to the student who teaches himself from books, correspondence courses, or from the hard lessons of practical experience as an apprentice or design office assistant. The reality factors in academic life may appear to derive less from studio project work, which is always imperfectly 'real', than from the discussion that surrounds such work, the exploration of angles of attack, and the slow take-up represented by the experience of community.

Design education *must,* by its nature, dig below the surface, and must at the outset be more concerned to clarify intentions than to get results. If it is sensible to see learning and understanding as rooted in the continuum of life, it may be that a really useful introductory course will only show its value in the full context of subsequent experience; i.e. several years afterwards. Conversely, an education that concentrates on short-term results may give a misleading sense of achievement and fail to provide an adequate foundation for subsequent growth. This is a thorny problem, because under the pressurized and success-conscious conditions in which we live, students are naturally anxious to prove themselves as rapidly as possible (to themselves and their contemporaries and teachers). Something as intangible as the growth of understanding may seem a poor substitute for the almost measurable achievement marked by a high output of design projects, however specious or thinly considered such projects may be.

Degree or diploma work is a serious business for students, and can cause a great deal of anxiety, particularly to those who have a (perhaps well-founded) dread of examinations in any form. If such students will realize that their school work is but an intensive phase of a very lengthy design education, perhaps they will feel less undermined by those pressures. The purpose of coming to a design school is *not,* primarily, to gain a first-class degree, but to make a constructive use of several years' education. Some students, on the other hand, will find the edgy business of first- and second-class degree a positive stimulus or a useful objective to work toward. This is a matter of individual psychology: in a good school the course work should be rewarding enough to keep such matters in a tolerably *Angst*-free perspective. An intelligent design course will also recognize that much design work is shared co-operative effort: students will be encouraged to help each other. In such matters it is sometimes necessary to educate the educators. A small-minded and up-tight view of human experience will take small differences of accomplishment very seriously. A more generous view will allow that we are all holding candles in the dark.

Teachers who are practising designers may also have problems, by coming to feel that they have 'lost' ideas that may have cost them years of work, given and assimilated instantly and almost unnoticeably in the ordinary way of teaching exchanges; only to meet years afterwards the reproach that they have exploited their students' ideas to their own advantage. As designers well know, there is a private notebook of the mind stocked with ideas in various stages of development, and it is these that seem to disappear into teaching situations – rarely, in the nature of things, emerging in the form to which the designer's greater experience would have brought them. The origin of these occurrences is soon forgotten, but if a designer is teaching for any long length of time, he may feel increasingly deprived of his own creative identity. Such difficulties become obsessive only in overstructured teaching and learning relationships, where the us-and-them structure seems to generate attitudes akin to property-hoarding. It may be said here that a good design teacher will try to help a student to think through problems at his own pace and at the level of his own attainment. Such a teacher will communicate both his areas of confidence and the limits of his own awareness, thus putting forward the fruits of experience in a spirit of positive uncertainty. In this discretion, and in this spirit, 'models' and 'constructs' may become acceptable and unrestricting – not to copy, but to examine the thinking behind them, and how this has worked out. Models may be verbal or material, and constructs will be artefacts of one sort or another. As to students and staff, Frederic Samson puts the matter pithily (addressing students at the Royal College of Art): 'the main difference between us is that your ignorance is superficial but mine is profound'.

Unfortunately such remarks would hardly be well taken in some schools and training colleges: too much is implied, of the life of knowledge, and of knowledge of life. Yet the challenge of a fast-changing global view of life stretches the imagination like elastic; it is ever more necessary that experiences nearer home are tethered to credibility, as distinct from pretension. There can be no doubt that many institutions have become seriously overstructured for the job they have to do. In education and elsewhere, the grip of outworn forms may discourage new energies beyond the toleration-threshold within which differences are normally met and resolved. The parties involved find themselves talking in different languages – in the same tongue, but with radically conflicting assumptions. Such new energies may be quite unformed, supported less by argument than by exploratory behaviour as such, and may be reaching toward new insights along unfamiliar paths of social exploration. This is the genesis of the student demonstration or sit-in.

Sit-ins and other demonstrations can become merely fashionable, or may become as rigidly institutionalized as the conditions they affect to challenge or displace. However, it is short-sighted to regard these happenings as mindlessly destructive, when they may generate and focus an extraordinary energy of self-education, quite remote from the ostensible aims of 'protest'. Indeed it is possible for those who detest revolutions to see such actions as genuinely an attempt at social adaptation; that is, to conditions which more ordinary channels of expression are failing either to contain or to anticipate. (Traditional forces of reaction will doubtless learn to keep up with the times.) Certainly a husk of discarded usages is revealed on these occasions, and one that has failed to grow with the perceptions of an impatient minority. At worst, the reaction provoked is too violently defensive to promise more than deep disillusion to the revolutionaries. At best, an institution is jerked out of complacency into a new consciousness of its privileges, and, for the participants, there may be a deeply moving experience of those twin principles for human conduct – solidarity and reciprocity (all for one and one for all, and an active spirit of empathy in human relations). The result is not (as commonly supposed) a developing group hysteria, but may be a notable growth of self-knowledge and individuation.

It is reasonable to mistrust any theory of conduct which draws life from the presence of an external enemy (or the necessary creation of such). Yet there is a dangerous doublethink that permits violence and apathy to be institutionalized in society, whilst deploring the faculty that young people have for detecting humbug in their elders. The intelligent way to meet any promise of new life must be with gratitude untinged by cynicism. Given favourable conditions for experiment, then the worth and staying power of new ideas will swiftly demonstrate themselves. It is only remarkable that *in practice,* official educational attitudes are so deeply resistant to change that in some places the waters have settled placidly over the disturbances of 1968 as though they never happened; change being too often substituted by expansion. Above all it is necessary to defend the notion, and the fact, of pluralism in education; against the ever-possible threat of some new, centrally directed, and paralysing orthodoxy. Taken as a prescriptive cure-all, with no awareness of local conditions, even a 'network' can become a cage of emptiness.

Formal education is now under attack for its own credibility. If the gap between technical capability and social imagination is to be bridged, or if that problem is even to be energetically appraised, education must lose much of its formality in pursuit of a new warmth and flexibility of outlook. This does not imply abandoning standards of intellectual discourse or of real attainment in any field: on the contrary, it is the only hope of their survival within the education system. Every single concept of structured education needs to be freshly seen and needs to be overhauled (and, where necessary, rebuilt) from first principles. Given the will, such principles are neither hard of access nor beyond the reach of agreement. The essential realization is that the informal truths are closest to sources of wisdom and creativity, and that experimentation at all levels is not a luxury, but a first need of this century: in our work, in our leisure, and in the precarious survival that we hope to make more secure for our children. In all such efforts, there will be failures again and again. It is important to recognize that there is an honourable difference between failure by default – doing nothing or not even recognizing what might be done – and failure in an effort to do something worthwhile. What we may conceive as a genuinely free school may be as elusive as a free society and prone to as many internal contradictions; yet freedom must remain the one permissible tyranny.

As for design, there are times when to say no is a constructive act; to say yes, *as a designer* looking to the future, is to join social commitment to a mastery of particulars. In education, all we can do is make good work possible, and be alert to its coming; never fooling ourselves that all good things come easily. To work well is to work with love. A hail of words, like rain in April, can do no more than keep the air sharp and sweet and the ground springy underfoot; and that is the best a formal design education can hope to do – relevantly.

4 Design education: practice (GB)

There has always been an apparent conflict between the training up of 'marketable skills' – a short-term social requirement to fill jobs – and the role of education in refreshing social values, calling jobs and indeed everything else into question, and thus helping to continuously redefine the longer-term social perspectives. At its best, a design education might claim to bridge this gap, by the occupation of what Bruce Archer has called a 'third area', a body of 'practical knowledge based upon sensibility, invention, validation and implementation' – the world of doing and making which is distinct from the worlds of the humanities and the sciences, and which embraces the useful arts as normally understood. There are difficulties in this position, but the aim is laudable and constructive, and gives a growing confidence of definition to design studies in the secondary school curriculum.

The wider problem is its own implementation. It is only necessary to read Alfred North Whitehead on *The aims of education* (1929) to realize that everything worth saying about education has been said many times over, often by men with a distinction of mind rarely to be found within the education system. It would be quite sufficient to refer interested students to the least technical of educationists (such as Whitehead and Martin Buber), for these men spoke from a rich and fertile sense of life, and are quite free of the pedantry and self-satisfaction that seems to afflict the specialists – *but* for that perennial mismatch of theory and practice that makes it necessary to speak of lesser matters; and somewhat in the style of a pamphleteer. It is not easy to write of practice in generalities; and without all the revealing detail of the particular case; but it is necessary to try.

Robert Sommer in *Tight spaces* mentions an American university in which (in the dormitory lounge) all the chairs, tables, and ash trays were chained to the floor. Institutional life has not gone quite so far in Britain, but wherever the forms and structures of education cease to embody its necessary aspiration, some very strange things are likely to occur. When they do, it is like the familiar signal from a design problem that is labouring unwarrantably in its detail; the signal tells us, with experience, that we must go right back to square one:

the root concept is wrong, or falsely interpreted, or in some way insufficient. Most of us now know (as is argued in part 3) that a design education is grown into over a working lifetime, and that colleges should take a modest, lively, and encouraging view of what they have to offer as a contribution to this journey. Yet the colleges persist in stressing short-term attainment of a kind that comes close to being not only falsely fair, but anti-educative in a demonstrable manner. It is well known and widely admitted that educational opportunity neéds to be serviced over a person's working lifetime (and beyond). Yet the skills and facilities that might benefit a local community are withdrawn progressively to service massively documented 'degree-level' courses, often of curious social and intellectual provenance. There are always reasons why buildings should be closed to all but the chosen (a matter to which Henry Morris must have been sensitive at Impington). That many students find themselves in the wrong place at the wrong time for the wrong reasons reflects the fact that instead of being given a life-ticket of educational availability, they have arrived at a certain point on a ladder where it was a matter of moving off, perhaps irretrievably, or moving on. These are commonplaces. They merely introduce the question – argued by Illich and others – as to whether the education industry should be flatly opposed, as being, at worst, an elaborately self-perpetuating fraud, and of its nature, increasingly anti-educative; or whether something can be done in a piecemeal way to improve matters. I do not propose to argue the case at length; it has been well ventilated (from both sides) elsewhere. The point is that the question can now be posed in these terms; and supportably. Nor is it necessary to say that the system is upside down in its priorities, nourishes distortions of intelligence, and deifies false knowledge – these are matters of intrinsic concern and debate – when at a structural level there are simpler matters worthy of remark.

It is possible, for instance, to argue that the whole structure of British art and design education (if not that of higher education generally) is observably standing on its head. Higher management, who not infrequently, may know the least and do the least in the specifics of design practice (they shuffle papers and 'speak to motions' in Committee), take home most of the pay, denuding resources where they are most needed by the students (for example, to pay part-time staff in the studios and workshops, who may be young practising designers). At the receiving end, those who most need education are the least likely to get it – they are 'unqualified'. If lucky, they will receive training. Thus differentials, so beloved of leaders the world over, are not only entrenched, but slowly widened. If this situation is immovable, as indeed it is, because the education industry looks after its own (cf. the 1980 Clegg Report on salaries), and nobody is of course dismissed except for serious misdemeanours, the educative *value* of such a system might seem questionable on these grounds alone.

As is well known, in a Moneyman and media-dominated society the problem of credibility becomes ever more urgent and explosive. At the simplest level of competitive self-interest (well below the normal preferred level of educational discussion), a rough justice will increasingly have to be seen to be done (because extremes of inequality are increasingly visible, and presented to people continuously in their own homes, a special kind of affront). Yet nowhere does the credibility gap seem more manifest than in the social distinctions that the education system upholds, and silently perpetuates. It does not directly follow, but it is an associated failing, due to corrosions of good faith, that what is generally said splits away retrogressively from what in fact occurs, 'is the case'; the final and intimate educational encounter being, on the one hand, faultily educative, and on the other, only too revealing of those concealments that life is thus shown 'to be about'. It takes no more than a few swift strokes with the brush to recognize an outline. If this is a realistic introduction to the transactional world for young people, can it really be the measure of our educational ideals and responsibilities?

Let us suppose – it can happen – that a senior educationist is propped up by his over-large salary, the status-assurance of his position, and the length of his holidays, but that otherwise he is beyond making an active or informed contribution. This is a difficult human problem in any area of management; the more so if it becomes unmentionable, and if resources are seriously short. If, however, his staff are enjoying a short-houred four-day week, the vicarious life (with or without sexual overtones) derived from the company of young people, and work that falls comfortably within their capability rather than stretching it, then there is every temptation to look the other way and join a staff-student connivance (see part 3). That all this is actually bad for work and morale in the long run, is easy to overlook. In educational terms, such problems are potentially grist to the mill, because they speak of human frailty – something we all share – rather than any conspiracy to defraud; but only if their implications are admitted into the world of educational discourse. The same applies to the desperate and sometimes pathetic scramble to upgrade vocational courses to 'degree-level' and then to proliferate such courses laterally; if this can be justified at all, when resources are so limited in primary and secondary education (and in the smaller colleges), this form of empire building must at least be seen in terms of human folly and weakness, rather than human wickedness. In art and design, the smaller independent schools, often with good traditions of personal relationships and work, may be threatened with 'rationalization' continuously; much to the cost of their forward planning. However, in all schools, small or large, questions must be asked and be seen to be asked, because it is of the nature of an education to question itself, or forever hold its silence.

As to differentials, this is always a touchy subject. In art and design, the half-spoken fear that 'we may lose our best men to industry' is usually so far from the truth that nobody with a sense of humour will trust themselves to utter it. It might well be added that if differentials are to obtain at all, and if economic self-interest is what makes the world go round, etc, etc, then presumably those who teach the most on the floor, whether in primary, secondary, or higher education, should be given the incentive to stay there rather than opting out of the hard graft into management. Similarly the most privileged senior tutors and administrators might be awarded rather a small salary, in keeping with the high moral tone of their frequent pronouncements, and the extent to which they are looked up to . . . After all, Moneyman is not an educational model; he is a creature of aggressive salesmanship. Few would equate the two.

Unfortunately every breakdown of interior life in education, of organic content (matters discussed in part 3) is of course compensated by the scale and pretension of educational institutions. Thus the difficulty becomes self-reinforcing. The unit of confident informality becomes larger, Colleges merge into Polytechnics, Diplomas become Degrees, paperwork becomes increasingly mountainous, meetings and committees proliferate, false anxieties are overlaid upon perfectly real ones, all the doors and drawers and cupboards suddenly seem to be locked when students need them open, notices appear in triplicate about safety regulations, children and animals are banned from the building . . . It is at this time, of course, when the simplest tasks seem to be formidably difficult, or curiously empty, that the prospectus statements truly resonate and come into their own.

There are serious reasons in our present way of life, in the whole prospect that confronts students, to ask of education that it become a world of open discourse, that it declare its hand; that every effort is *seen* to be made, in the puncturing of pretence, in the fostering of authentic relationships, in the talking-out and the sharing of responsibilities, in the difficult pursuit of concrete instances for general principles. If the previous paragraphs have strayed into a world of caricature, it is a world the outlines of which *too many* will recognize, as they did in 1968. It would be good to think that the argument was pushing, in all cases, against an open door. Knowing perfectly well that many of the difficulties are deeply rooted in the anxieties and frustrations of our age, are there yet simple things we might do to improve the chances for design students?

First it must be said that an ounce of spirit and initiative is worth a hundredweight of correctly worded course documentation. Educators should earn their salaries by helping to gladden the hearts of their students (and perhaps, to begin with, their own). Disciplines of the mind will follow. This means making institutions, by whatever means, into communities. A community is a place where people belong, in which they can put down roots, and in which they can grow their experiences harmoniously. Growth will never exclude conflict or unhappiness, but a purposeful community should admit of such root experiences, and allow for their effects. If this is too much to ask, life being what it is – then let it be an encampment, but at least have people talking freely to each other, and a minimum of decision-taking behind closed doors. Role-distancing, as one of the grosser forms of status-assurance, is really not on in the second half of the twentieth century. Educationists should not need someone like Donald Schon to talk about the inefficacy of centre-periphery models. It is well enough known that the only acceptable forms of leadership bespeak quality of involvement and a well-understood reciprocity of function in a group. A worker at Lucas Aerospace is quoted as saying that 'management is not a skill or a craft or a profession but a command relationship; a sort of bad habit inherited from the army and the church'. There is no excuse (except that of ignorance) for higher education being open to such a charge. In every country of the world it might be added : we learn to live with Moneyman and Trend, but God preserve us from the aparatchiks. They take an awful lot of loving.

As for encampments, John Andrew Rice (of Black Mountain) remarked that 'colleges should be in tents and when they fold, they fold'. This is certainly too much to ask. It is not too much to expect, however, that senior educationists should know who John Andrew Rice *was,* and the nature of his difficulties . . .

Higher management (with which it seems we are stuck, and let us hope in most cases, blessed) should set an example not only in self-education, but in propounding and defending the most open and demanding standards of critical debate. It is absurd that educators can still avoid declaring their credentials; the more so, when funds are so short. 'Course validators' from whatever source should always be expected to state their own critical position before the assembled students and staff, and desirably they should also show something of their own work. If they can do neither, the least that can be expected is a graceful apology. The same goes for the appointment of senior staff. Not all teachers practise their own subject actively, nor should this be expected of them. The aptitudes and the resource required of a good teacher may not be those of a practising designer. On the other hand, students are entitled to expect that their teachers are speaking from personal experience. If a design competence is claimed, it should be made manifest; and it is hard to imagine a useful teacher of design without some coherent critical position to put forward. These simple exercises in communication would not only be stimulating and educative for all concerned but might helpfully sort out the sheep from the goats. This is not an examination system in reverse, but simply follows from the view, upheld in this book, that open education is now even more necessary than open government. If 'free men govern themselves', then let education be the nursery of responsibility, which can only be nurtured in a world of open discourse and credible behaviour. Again, if *all* design teaching appointments were fixed at a three-day maximum, or even half a permanent full-time appointment, staff would actually be obliged to do their own design work, instead of complaining continuously about the lack of opportunity, or losing touch with it by default. This could bring new faces into the schools, tone up morale, and induce realism.

If numbers will stand it, there is much to be said for a shared appointment between a director of studies as head of school, with full academic responsibility, and a convenor of studies, who handles administration in all forms but also chairs meetings and lectures, receives visitors, and in all such matters formally 'heads' the school. The academic director is thus freed to programme his work very flexibly, and to stay with the students. In this arrangement the convenor is permanent, full-time; the director has a five-year (possibly renewable) appointment, with flexible hours to cope with the expected continuation of his or her design practice. The convenor 'convenes' people, time, place, and occasion; the director is concerned with the content of studies, their aim and their direction. At present, of course, the designation 'director of studies' usually means something quite different, in keeping with a general tendency in higher education to tell-it-how-it's-not; the director being an administrator who does not direct studies. It must be stressed however that such arrangements call for a positive and forward attitude from both parties, and an equal sense of involvement in the life of the school.

If 'difficulties' are approached in a negative spirit, a school will lose much of the fun, enterprise, and commitment that might derive from lively relations with the local community. In every city there are projects – schools, hospitals, play groups, welfare organizations – that are often desperate for help and equipment. The conditions for 'success' are not nearly so critical as in ordinary commercial work – except in the case of special purpose equipment (say) for hospitals that demands a research programme. Even here the design schools should be making a contribution. Every school should have its own design office in which teaching staff are kept creatively active for a part of their (paid) teaching time, and in which students can work – in effect – as apprentices, during a part of their design course. This is a very fast and practical way of learning. Projects in which students work both as designers and executants should be chosen with some care, because students usually underestimate the time spent in supervising and completing a job – something, on the other hand, that they might well discover for themselves. Schools might also run a local shop; indeed, there is no limit to the ideas that might emerge from a socially committed view of education, as distinct from the view which directs effort toward window-dressing and constant onerous assessments of a student's supposed 'progress'.

It seems unlikely that design and architectural studies will integrate to any useful effect: this could only happen through the development of an RIBA two-tier qualification for architects, for which a non-specialized design course would provide a foundation in common with other intending specialists (see part 26). At least it would be pleasing to think that architects might be eased a little in their procrustean beds, to allow a slightly more lenient view of what an architect might usefully be and do. That architectural students do often enough repudiate their architectural identity and heritage (temporarily), is not quite the same thing. Another overdue experiment, the organization of a design course largely by women for women, to see how far the specific content of modern design has become male-dominant, seems at present not in view. Certainly it would seem reasonable in design schools to work toward a proportion of female teaching staff comparable with that of female students in the school.

Finally, it is to be hoped that the national validating bodies will abandon their 'you scratch my back, I'll scratch yours' procedures, to a principle of full academic autonomy. If a teacher does not know what to teach or how to teach it, he or she will be little use to the students or themselves, but to have an opposite number from another college telling him, is merely to compound a folly. If on the other hand the two are nudging each other and comparing notes on hotel accommodation, the procedure becomes ridiculous. Nor should the possibility be overlooked, that the worst teachers and the worst courses may be the best at selling themselves. The necessary effect of this rigmarole is the encouragement of timidity and conformity in course planning (or more seriously, in the whole educative stance of a school), and of surface effects and their false standard of attainment, which may have many anti-educative implications (see part 3). There is also a fearful cost in wasted time, window dressing, salaries, travel and hotel expenses, and paperwork.

With due respect to British institutions like the CNAA (and there are no personal allusions in these remarks), it must be said that there is a further difficulty: an inbuilt tendency to false omniscience. That public institutions can become too big for their boots is a well-known international problem, and a perennial one; as the students and dissident staff felt obliged to demonstrate in 1968.

It is good for schools to have their working position open to scrutiny and to challenge (from any source); this could as usefully come from trade unions, local industry, consumer groups, as from schools elsewhere in the country. A school should at all times (collectively) have a notion of what it is doing and why; and such notions should be defensible. Schools should thus be exposed to criticism, to influence, and even to pressure; but their ultimate autonomy of decision must be respected. It is a fallacy to suppose that Big Brother, or his accredited representatives, must know better. Such people do not exist; God does not personally select and appoint them. On the contrary, when encountered face-to-face they are as likely to be embarrassingly ignorant and parochial in their attitudes, as well-informed or imaginative. As we have been warned by Erich Fromm and other wise men, it is the former category who come to dominate in the corridors of validation. This may be compensatory to unfulfilled lives, as happens to all of us; or more simply, it might be noted that those who enjoy wielding power and influence unobtrusively, being thus drawn to social structures that permit this to occur, are not always persons of imaginative ability. The power of all such bodies should be prudently held back to an advisory, consultative, and informative function.

For all these matters, and many more, the important and educative distinction is between attitudes of bureaucratic correctness, and those of informal exploration – a distinction also between putting down the deviant as a source of 'trouble', and encouraging deviant energies as one necessary source of hope for change and survival. Opposed attitudes here see the same bottle half-empty or half-full. On one view standards of aspiration will be lowered sufficiently to ensure success (and as Lethaby remarks 'the happy mean is likely to be the meanly happy'). On the other, every possible way is sought to give practical help and encouragement where fresh approaches are in need of support – even where they might seem, by existing criteria, to be courting failure. This situation is discussed in part 3; it is one in which the possible destruction of the planet is a matter, not for nightmare or fantasy, but for sober predication (or even bored acceptance). Perspectives must, surely, adjust a little . . .

The Black Paper people (for example, G.H.Bantock and Rhodes Boyson) look to discipline and a revival of the grammar schools for their vision of educational excellence. It might be urged that effective culture transmission has more to do with credibility than with imposed discipline at any level of schooling. However, they might ask, from their right-wing perspective, how it is that the word 'disinterested' ('not deriving personal advantage') is rapidly coming to mean *un*interested (a usage 'revived from obsolescence' – Chambers); and just what this loss of meaning points to, for the survival of excellence in our culture. The unholy marriage of Instant Trend and Moneyman is surely an achievement of their own social philosophy.

Such critics also suppose that the progressives have always one hand itching towards their Dewey and Rousseau, where formerly it was reaching (more properly) for the rule book and the cane. This is a black herring. The damage done by schooling, where it appears in higher and post-graduate education and is identifiable as such, is never (in my experience) remotely traceable to the poison of progressivism in primary school education. On the contrary, it proceeds from the failure to educate for the conditions of this century, the common philistinism and mediocrity of standards, the rigidity of an examination-obsessed system that excludes, or diminishes, the best things of life – music, dance, drama, poetry, art, design, construction – into the realm of 'extras'; and lastly, it proceeds from a real failure to pursue excellence at all, unless excellence of moral concern, of sensibility, is filleted out of it. (The result becomes examinable.)

Fortunately there are always devoted and warm-hearted and skilful teachers in the worst of schools (whether of the left or the right) and the system itself has yielded, at secondary level, to an enlarging view of what design has to offer, even as an intellectual and a subject-coordinating activity. It is also amazing what some young people do survive. However, at the start of this discussion, its burden was referred to as a 'lesser matter'. This is true simply for the concern of this book, which is good work and the possibility of good work. Self-education, in and out of the schools and colleges, becomes a search for valid traditions in which to place ourselves, usually at a very minor and very humble level of accomplishment. It is something if such traditions can be identified, honoured, and explored a little in the schools.

What can the students do about it, apart from getting on with the job, and perhaps raising consciousness of their situation? I say 'perhaps' not out of pessimism on this score, but to qualify what is dangerously near to a cliché: 'raising consciousness' does not mean its substitution by Marxian dogma. Students should certainly be aware of the legacy of competitive attitudes with which they enter higher education, and which makes co-operation so very hard to achieve. They might view with a proper scepticism – and a little charity – the self-importance of the education industry, and the earnestness of its self-regard. It is hard to *know* that education takes place largely outside the schools, but students can make intelligent guesses on the basis of the evidence, as we all do. Lastly, they might honour the situation of privilege they share with their tutors, and spare a thought for equally deserving young people without the benefit of even a questionable higher education.

Here is A.N.Whitehead on the universities:

'A university is imaginative or it is nothing – at least nothing useful . . . The whole art in the organization of a university is the provision of a faculty whose learning is lighted up with imagination . . . Education is discipline for the adventure of life; research is intellectual adventure; and the universities should be homes of adventure shared in common by young and old.'

5 What is good design?

The 'goodness' or 'rightness' of a design cannot easily be estimated outside a knowledge of its purpose, and sometimes also of its circumstantial background. This is no reason for timidity of judgement; a man must reserve his right to say 'I like that; to me it is beautiful and satisfying, and more so than that one over there that works so much better' – or, 'this is a good workmanlike solution, thank God it has no pretensions to Art'. Theoretically, a well-integrated design should come so naturally to eye and hand that neither of these comments will be called upon, but human nature isn't so simply natural and nor is human society. An optimum solution is possible where the conditions for verification can refer to absolutes; a daunting and illusory requirement in most design situations. On the other hand, a design can say to us 'here is a problem that is so well understood that it can be felt to be moving toward an optimum solution; the design is inclined in that direction'. This is designers' talk; the user of a product will not be too interested in the skill with which a designer has met his constraints. If a design is so well wrought that overtones of meaning are present, so that the work can be experienced (optionally) at many different levels simultaneously, then it is a condition of *organic* design that the further harmonics must not clutter or deform a simple level of acceptability.

For the designer, good design is the generous and pertinent response to the full context of a design opportunity, whether large or small, and the quality of the outcome resides in a close and truthful correspondence between form and meaning. The meaning of a good garden spade is seen in its behaviour, that it performs well; in its look and feel, its strength and required durability; in a directness of address through the simple expression of its function. More complex objects, places, equipment, situations, may well exhibit less obvious dimensions of meaning – of which one may be the property of reference discussed elsewhere in this book. A design *decision* may prefer some determinant principle of action to a material outcome. As a social activity, the integrity of design work proceeds from the understanding that every decision by one human being on behalf of another has an implicit cultural history. Design is a field of concern, response, and enquiry, as often as decision and consequence. In this sense (also), good design can both do its job well and speak to us.

Every design product has two missing factors which give substance to abstraction: realization and use. These are the ghostly but intractable realities never to be forgotten when sitting at a drawing board. In a similar way, any discussion of design philosophy must never stray too far from nuts and bolts and catalogues and every kind of material exigency: a designer breathes life into these things by the quality of his decision-making. Thus his concern is truly 'the place of value in a world of facts' (see the book by Köhler of that name) and the outcome can (or should) be a form of discourse; but not a verbal one. His work can be said to deploy the resources of a language and be accessible to understanding through the non-verbal equivalents of intention, tone, sense, and structure, (cf. I.A.Richards), but there are other and more directly functional levels of experience which – as has been said – must come to the hand with all the attributes of immediacy. Most of the time a designer finds it hard enough to do small things well. Any number of broader considerations must not distract him from that task, but rather enliven and give sanction to its meaning.

A product must not only be capable of realization through manufacture, but in its very nature must respect all the human and economic constraints that surround production and effective distribution. This may seem obvious in the case of product and communication design. Similarly, it is difficult properly to evaluate a building without some idea of the cost factors and the client's briefing. Difficult, but not impossible; because a clear design will generally manage to state its own terms of reference, unless disaster has intervened at some stage to distort the central intention of the work. There are many cases in which a good design will be discarded for reasons which seem arbitrary, perhaps to be replaced by some meretricious product with a better sales-potential. Again, a perfectly adequate design solution, the result of much care and imagination in its development, may never reach the public at all. In this respect the artisan designer may enjoy a freedom denied to the designer for mass production (though his economic problems will limit the scale of the work), and much experimentation in form-giving will necessarily occur in situations exempt from marketing difficulties. These will include one-off jobs, limited production runs, and public work (schools, hospitals, airports, for example). Much early discussion in the modern movement assumed – broadly for social reasons – that product design (mass production for the consumer market) was the centre of inertia that had to be revitalized: Herbert Read's book *Art and industry* reflected this assumption in the 1930s. Gropius's *New architecture and the Bauhaus* contained a classical statement which seemed to imply that product design would move inexorably toward the 'type-form' for the problem examined.

As things have turned out, the most interesting work has happened, of course, where it was economically possible. The domestic consumer market has gained an important component in DIY (do-it-yourself) which in itself demands a reappraisal of the designer's role in the areas affected. The mass production of building or service components (such as pressed-out or moulded bathroom service units) has hardly approached the potential seen for it fifty years ago. The notion of *place* as the focus for communal achievement has scarcely fought off the demands of *occasion* and mobility, despite moving and articulate pleas from Aldo van Eyck and others – and despite the continuing reality of place as a factor of ordinary experience, eroded as it is by communications, and the rarity of imaginative work in this field. It is a mistake to see the designer's work as conjuring up new worlds at the scratch of a drawing pen: there are many fields in which the designer could profitably work with (for instance) do-it-yourself and co-operative housing agencies; and there are fields in which a designer can and should respect the organic continuity that surrounds people's lives. Two examples: the interior designer is doubtful of his 'responsibility' because every one knows that architects should design their buildings from the inside outwards. In fact, there are plenty of buildings that are simply weather-proofed and service-provided shells, waiting for specific uses to be provided for. However, leaving that aside, it is not a necessary argument to suppose that – given adequate social resources – the whole of our physical environment should be uprooted and totally replaced at regular intervals. This is a dangerous fantasy. In fact there is plenty of scope for the adaptation of existing buildings to new uses (a so-called slum area is as much a pattern of relationships as of decayed buildings) and this is interpretive work for which the 'interior designer' could be well-fitted. Again, there are plenty of structurally sound buildings that could be given extended life with the aid of a loan and a do-it-yourself handbook. As it is, the lunacy of high-rise development has only recently been seriously questioned; in practice, land values give rise to extraordinary palaces for paperwork springing out of areas of private squalor, and the simple things – like the provision of neighbourhood amenities – are neglected in favour of drawing-board schemes which may seriously debilitate the life that 'squalor' sometimes reflects. A run-down neighbourhood may need a lot of things but the problem must be seen in more than a tidy-minded way; every problem, however complicated by planning and growth statistics, is met with concealed assumptions (and often concealed economics). Here is ground both for humility and for diagnostic sensitivity in the way a designer approaches his work.

The difficulties for product designers are not just a matter of plain villainy on the part of manufacturers; they are, in part, a consequence of capitalism. Whilst strange things do go on in boardrooms, it must be realized that a well-designed product must be sold competitively. Experimental work may be chancy as a sales proposition. As things are, a first duty of a company director is to make his company profitable (which he may conceive as a first duty to his shareholders), and a second duty is to keep his work-people in continuous employment. Experiment becomes a closely calculated risk, very much at the mercy of the buyers in the retail trades, and subsequently dependent on a successful advertising policy, public response, and many other factors. In the furniture trade there are a few companies who have tried to maintain reasonable design standards, against the hope of improving them as the market 'softens' sufficiently to warrant further advancement. Such companies have relied on contract work – furniture for public buildings specified by architects or local authorities – to help carry them forward. It will be seen, at least, that under ordinary production conditions, product design cannot easily be evaluated against absolute standards, yet products meet constant criticism on such terms.

Unfortunately, it is also true that there are innumerable products that are just very poor realizations of a straightforward and entirely non-experimental design concept. They could have been marketed just as easily had they been designed with more distinction. The design capability simply was not there. Designers should be aware of property-relations as a conditioning factor in the way they design (and think about design), but no designer should fool himself that given 'a better society' it would then be magically easy to design *well*. A designer who stops designing in the hope of better things may lose his ability to design anything at all; to this extent people become what they do. Here, an idealistic student might consider the partial truth in the saying 'a few are artists, the rest earn a living' – which in caricature might be said of every profession and not less of the sciences than the arts. Those who elect to put their work before everything else, which is merely one of the conditions for complete mastery in any field, must fairly expect life to present some difficulties.

The hard facts of a market economy are easy to overlook in the relatively permissive ambience of the average art or design school. Although academic life is subject to its own peculiar stresses, economic sanctions are not pre-eminent among them. Fortunately, students need not harden themselves against a perpetual winter of creative frustrations: the situation is not as depressing as some of these remarks might suggest. It is true that a designer's freedom will reflect in large measure the values of the society in which he works. Designers are not privileged to opt out of the conditions of their culture, but *are* privileged to do something about it. The designer's training equips him to act for the community, as (in limited respects) the trained eyes and hands and consciousness of that community – not in some superior human capacity, but in virtue of the perceptions which he inherits from the past, embodies in the present, and carries forward into the future. He is of and for the people; and for them, and for himself, he must work at the limit of what he sees to be good. The sentimentality of talking down, or working down, is a waste of the social energies invested in his training: thus can 'social realism' enshrine the second-rate.

If society is geared to satisfactions on the cheap, the designer has a special responsibility to straighten himself out in that respect; to decide where he stands. When real needs are neglected, and artificial ones everywhere stimulated into an avid hunger for novelty, sensation, and status-appeal, largely (but not wholly) for reasons of private or public profit; then here is his own nature, his own society. He is involved, and he must decide how best to act. It should not surprise him to find a thin and pretentious reality informing the design language of the world which he inherits. A Marxist (or anarchist) analysis may be one tool to help him sort this out, but he will hardly need to put on Marxist spectacles to see that a veneer of good taste has 'reference' to certain obvious social conditions and is not the whole of good design. The design student may sometimes find that the industrial scrap-heaps, the surplus stores, and the products of straightforward engineering, will yield images of greater vitality than will be found in more fashionable quarters (though even here, fashion spies out the land). Such a situation is a challenge, and as such must be studied and understood.

Yet it is still no answer to live in the future; every skill must be nurtured by a commitment in depth to the present. The meaning of creativity may be seen as an equation which resolves this apparent paradox. Work that lives is rooted in the conditions of its time, but such conditions include awareness, dreams, and aspirations, as much as the resources of a specific technology : such work respects the past and actually creates the future. These problems, and their wider implications for human happiness, will necessarily concern students of design, because no one can make truly creative decisions without understanding; and without a real participation in the constructive spirit of his time. *This spirit must be sought out,* not necessarily by intellectual means, to be honoured wherever it is found.

Those who are depressed by the shoddiness of our environment (except in those areas of economic privilege where it is customary to buy up the past), should study the spirit of the modern movement in its development from the turn of the century to the late 1930s. Here they will find themselves in good and most various company. As Walter Gropius often explained, the modern movement was not some matter of dogma, fashion, or taste, but a profoundly wide-ranging attempt to encompass the nature of our twentieth-century experience and to meet its physical demands with a constructive response.

What may excite us mostly about this phenomenon is its surface appeal, the tangible achievement; a whole world of very explicit imagery conjured, as it must now seem, out of nothing – an entirely fresh start. The fact of conditions historically different from our own does not diminish the marvel of this achievement and its continuing relevance. This has nothing to do with imitating the forms of the past (near or distant) : anyone who sees the modern movement in stylistic terms will fail to understand its radical nature. It is also necessary to accept that most work of today (meaning 'modern' or 'contemporary') is an enfeebled and misunderstood derivation from this earlier work, almost wholly removed from its inspiration, its most deeply rooted concerns, and the force of its guiding spirit.

The effort of zero, of the *tabula rasa,* of the new beginning, is not in principle a stylistic option (though in retrospect it may be so viewed);

it is an effort consequent upon certain perceptions, for which, obviously, there will be equivalents in prophetic or diagnostic acumen, across a whole civilization experiencing radical change – or perhaps it is truer to say, waking up to a foreshortened view of what such change might seem to imply. Earlier models, and the canons of *idle* change, the sports of fancy, become suddenly and drastically inadequate. The call is certainly to 'clear from the head the masses of impressive rubbish' and to 'make action urgent and its nature clear'; it is also to 'look shining at / New styles of architecture, a change of heart' (the quotes are from the poet Auden of the 1930s) but most of all it is an effort of address toward the irreducible; that modest yet most demanding of entitlements. Is there indeed any alternative? – but silence; as George Steiner has remarked in another context. The possibly prophetic nature of such insights is often overlooked, especially when accusing the modern movement of a false and shallow optimism; as though a culture of utility could not be expected to ask 'what is it decently possible to assert, given the claustrophic banality of a present, and the seeming threat of a future?' Less, perhaps, an insight than an indistinct awareness, and one of two negative imperatives at work, the other being *the need to stop telling lies.*

What more active principles are involved (or indeed, derived). It is usual to account for the modern-ness of modern work in terms of influences and precedents, the technological and social pressures, new materials and techniques, the convergent history of ideas, and so forth. It is obvious that orders of form, and forms of order, are design specifics in a practical way and experienced concretely, not as a set of abstracted verbal propositions. Yet as Viktor Frankl points out, in his book written from direct experience of the concentration camps, it is curiously easy to overlook, in any analysis of human motivation, the tenacious strength of the human search for meaning. It is here that the irreducible, the without-which-not, the minimal, the verification principle, 'truth to materials', and the notion of accountability, have their roots. At a different level, it is fairly obvious that failing the imitation of natural form (the dead-end of art nouveau), a verification principle would move towards number and geometry stripped of backward reference or depleted symbolism. 'Clear expression', no unseen props, and the most for the least (the strengthening of signal and the reduction of noise) have to be seen as necessary correlates to any search from zero for significant form – given a situation of survival, as distinct from options consulted on a broad wave of optimism concerning human progress.

However, it was the second broad outreach of the modern movement, involving nine more positive principles as guides to action (in fact there are a few more), that rescued a search for meaning from being merely the celebration of a rather unattractive rectitude. These are the social principles – not unwarrantably, design being a transactional art – and they prefigure certain changes in human relations which have not occurred, but may well have to if our society is to become less death-orientated. It is in this sense that the modern movement might be said to be prefigurative, and in this sense that its effort was betrayed (forgotten) in the take-over by the complicated apparatus of commercialism. These principles may be briefly stated as follows.

The first principle is that of self-determination: the search for a sub-set of self-generative principles within the situation as found; as expressed by the saying 'a well stated problem is more than half solved' and, as suggested elsewhere in this book, 'a designer transforms constraint into opportunity'. As things get under way, the principle enables the job to speak up for itself with increasing confidence and fluency. Sometimes this is falsely seen, so that in fact a designer able enough to work in this way is not actually being instructed by the job, so much as providing a good fit (i.e. just being a good designer); the imposition of *arbitrary* form always throws up a lot of noise and is in other ways more conspicuous. However, it is interesting that this principle, which applies very well in a straight designer-client relationship, is also flexible enough to accommodate quite different design attitudes, including those that might be thought anti-design by those who see the modern movement in formal terms (see part 8 for a discussion of this matter).

The second principle is that of reasonable assent; that what is done should be essentially coherent, intelligible, and open to discussion. (There are problems of language here which will not detain us: from the standpoint of principle, which becomes 'generative' to the way something is done, the fuel of continuing intention is more important than the ash of dead fires.)

The third is that every part in a job should work for its living ('From each according . . .'). This implies distinction and emphasis not from 'privilege', or prior status, but from functional differentiation within the whole. This principle (aided by others) entails asymmetry, a clear structure, and certain negative imperatives mentioned earlier. (For example, in a chair, the absence of glued blocks reinforcing an unsound structural principle, or, alternatively, a structure no longer covert, but derived from glued blocks.)

The fourth principle is that objects should be designed as well as possible for use and not for profit; and that where an object cannot be designed at all unless by definition it is profitable, then the resulting compromise is against principle and not with it. The modern movement has (rightly) been accused of political naïvety by supposing that optimized design was conceivable in mass production, where marketing arrangements will ensure a good product being swept away to stimulate fresh demand. Anyone who wishes to see such matters imaginatively explored should read E.C.Large's novel *Sugar in the air*.

The fifth principle is that of anonymity; poignantly expressed in the effort to mass-produce objects of quality at low prices; more adventitiously, and sometimes trendily, in the way that designers like the idea of their Thonet chairs, jeans, clothes-pegs – not to mention the universal boot. The requirement in either case is that the particular should be seen as a special case of an *available* universal. The principle is also, implicitly, an attack upon the art-object so constituted by its scarcity value (for discussions of which, see John Berger's writings). An extended requirement is the suppression of unwarranted detail (which in turn entails a special case of the third principle, namely that elements are distinguished from components), such that the fuss of idiosyncrasy slips below conscious regard. The human being is thus freed to enjoy a 'true' idiosyncrasy, supposedly more authentic in being less object-fixated; so that what was once merely idiosyncratic becomes genuinely individual. Thus by discovering what is uniquely true to himself (as distinct from conferred status) the road is open to self-transcendence. (See Buber's *The way of man* for a poetic uncovering of this theme.) At a less difficult level, the principle finds expression in an allegiance to the 'set' or series (of knives, of wine glasses, or whatever) in preference to the unique and single object. The attentive reader will have noted that this principle is the most vulnerable, the most open to corruption, and the most liable to misunderstanding (old ladies deprived of their tea-cosies and sentimental possessions, reds under the bed).

The sixth principle expresses a deep desire for a new vernacular (grown out of the alienation we all feel) – seeking articulation in a popular, indigenous, locally based, and relatively unselfconscious design language; adding a sense of place to that of space, of repose and location as a counter to mobility, and so on. Less wistfully, the principle develops an interest in the simple and functionally-derived design solution, often with engineering overtones (cf. canal buildings and structures, barns, windmills, and small houses everywhere pre-1850). Local variants are usually involved – of materials and technique – and there is a predilection for small or controllable human scale. This principle has always run strongly as a current of inspiration in the modern movement, but in the early years found expression as its apparent opposite (i.e. as a paradox), namely, as an implacable resistance to sentimental craft-revivals and every other evidence of a falsely imposed vernacular: the conviction that there are no short cuts to Elysium except through 'the assimilated lessons of the machine' (a view characteristically developed by Lewis Mumford in *Technics and civilization*). The way back is seen to be through faith and through the wilderness, and a refusal to be conned by snap answers (e.g. the candles and leathers syndrome). It is interesting, of course, that some of this antipathy can be traced to the parentage of the modern movement in the arts and crafts, and that now the movement is finding it easier to come to terms with its parents in the realm of the Alternative (see part 7), the principle is gaining confidence; it is no longer so wilfully 'protecting the lost wisdom of the tribe' behind the fortifications of an arrogant modernism. However, it should be clear how the modern movement has come to be misunderstood on this point. Who was it who said that the true romantics of this age would be its most ardent classicists? I shall not waste time on the ignorant supposition that the modern movement is about high-rise buildings (or ever was, intrinsically).

The seventh principle is possibly the most important (though this is too puffed-out a word – formative is better) and it is the most closely linked to insights available from other areas of our culture: it is, of course, the search for relationship as distinct from self-sufficiency, or self-containedness, and everything that this might be held to imply at every level of decision in design. This includes, and perhaps most prominently, the complex realm of formal relationships, and how – once this principle is grasped – an entirely new way of working is disconcertingly revealed. If you had to explain to your aunt just what it is that makes 'typically modern design' different in kind to any other, this is the single principle that you would have to invoke. It is thus at once a principle of search, of reference, and of explication. Something of this is discussed in part 6 and elsewhere.

The eighth is the existential principle: that *there shall be nothing else*; and that what there is, shall be contingently respected. This most elusive of principles, is at the same time the most down-to-earth. From its employ springs the *ad hoc,* the improvized, the anti-institutionalized; and on the other hand, a healthy disrespect for the tyranny of absolutes. The joker in the pack.

Finally, the ninth principle (appropriately, that of the dance) is the translation of mass into energy and relationship. It is, in a sense, the dancing out of the seventh. This principle does not go as far as Proudhon ('property is theft') though it shakes the old boy warmly by the hand somewhere along the line; nor does it foolishly believe that the silicon chip will free mankind of its material adhesions; it is, however, the energetic principle, and as such must be assumed to be always embryonic with the hope of new life. It is also, of course, if you care to follow through Illich's indications, a holy celebrant. There are three lesser, facilitating principles – including that concerned with standard and standardization – but they will not be discussed here.

Now; if 'design' is overlooked and the preceding paragraphs reconsidered as metaphor, it should be clear that the nine principles translate very potently into the prerequisites for an a-political social revolution, and that this is no accident. It might also follow that a constructive art – design – can have, does have, an intelligibly expressive content. Of the two sets of ideas here, the first of which I described as 'negative imperatives', it could be said that the first collapses into a single pinched-face-personal-probity-principle, doomed to emaciation and 'the distortions of ingrown virginity' without the saving social outreach of the nine, which (on this analogy) are self-transcending. Returning now to the modern movement, the correspondence should be clear. The principles are, of course, interdependent (and much simplified) – as in the theory of compass adjustment, you have to put everything back together again before it makes sense. However, it would be inadmissible to have this discussion at all, were it not for the fact that for every principle mentioned here (including the first sub-set) there are *precise physical correlates* to be found, in the field both of object design and of design procedure generally. The notion that the modern movement prefigured certain qualitative changes in our society whereby human survival might be the better assured (under the industrial challenge) is not, on this argument, as fanciful as it might otherwise appear. It is true that this book is written from a declared standpoint, and therefore discusses design (here) in an unorthodox way. It is also true, and one function of the discussion to demonstrate, that this movement can never usefully be seen as a fashionable option that now happens to be *passé*. It is a serious demand upon intellectual assent and practical action.

Something of this is well expressed by Paul Schuitema, the graphic designer who worked in the Netherlands and Germany in the 1920s and 1930s.

'. . . We didn't see our work as art; we didn't see our work as making beautiful things. We discovered that the romantic insights were lies; that the whole world was suffering from phraseology; that it was necessary to start at the beginning. Our research was directed to finding new ways, to establishing new insights – to find out the real characteristics of tools and creative media. Their strengths in communication – their real value. No pretence, no outward show. Therefore, when we had to construct a chair or a table, we wanted to start with the constructive possibilities of wood, iron, leather and so on; to deal with the real functions of a chair, a living room, a house, a city: social organization. The human functions. Therefore, we worked hand-in-hand with carpenters, architects, printers, and manufacturers.

To reduce chaos to order, to put order into things. To make things more clear, to understand the reasons. It was the result of social movement. It was not a fashion or a special view of art. We tried to establish our connection with the social situation in our work . . . The answer to our problems must be the questions: why? what for? how? and with what?'

The attractive qualities of this statement should not blind us to the fact that the modern movement has always been a minority struggle, carried on against a good deal of practical opposition, and, at best, a widely felt social indifference. At least in its early days the conflict was capable of clear definition.

It is necessary to stress some of the background considerations which prompted modern design into being, because it is too easy to study the designs that emerged as specially privileged historical monuments, whereas the spirit that conceived them is still alive and accessible to us. In forming our own criteria for 'good' design, we cannot, of course, escape the half-conscious assumptions which make us always the children of our own time, but we do well to remember that our own concerns are in some respects closer to the pre-war period than to the world of the 1950s. A whole complex of emergent ideas, values, and experimental work was traumatically cut short by the experience of fascism, the horrors of Auschwitz and Hiroshima, and by the slow aftermath of cultural assimilation. Not only were energies dispersed in a practical way, but their foundations were uprooted. The implicit philosophy which underpinned modern design was never very far from what is wearily referred to as 'a rational view of man's conduct': the hope and even the confidence that if technology could only be integrated into meaningful value-structures, a new and fruitful way of life lay open to man's willing acceptance. The last war brutally damaged that hope.

For those who see the world as essentially an arena of conflict between good and evil, however, there can surely be no doubt that the modern movement stands for clarity, sweetness, and light; order, relation, and harmony; made accessible through the only means that are fully credible to our experience of this century. To open up a new age of revival or pastiche (the weakest form of wit) is merely to admit defeat in this sector of creative possibilities. Defeat may seem inevitable, radical change may seem a prerequisite for confidence of any kind, perhaps our civilization really cannot survive on its own terms; either of these realizations (if accepted) carries a more appropriate response than backsliding into a weakly fashionable eclecticism.

It is certainly obvious that a rational view must examine motivational forces with a more intimate sense of their origins, and the cost of their frustration. In the design field it is not just a matter of exchanging affective imagery for austerities that are deemed to have had their day. Here pop must be distinguished from vernacular; the one being a brilliantly successful commercial racket and the other being an unavailable option except at a stylistic level (see Herbert Read's poem, part 8). Our civilization has refined many hells, but in the realm of voluntary servitude it would be hard to beat the inanity of Radio 1 ('the Happy Sound' – GB –) with perhaps a few television commercials thrown in for good measure. A language of gesture and exclamation tends always toward infantilism; a measure of its warmth but also of its inadequacy. If a new synthesis of thought and feeling is to be attempted, we must think and feel our way toward the place of design in a necessary context of social renewal. Nor must we forget that a warm heart and a rather special view of history do not make a designer. Designing is very specific; a cultivated understanding is no guarantee of a specific creativity. This is the individual problem and a central concern of this book. For the social task we have fresh evidence all the time of man's fallibility, of his deepening technological commitment; of the nature of affluence divorced from social or spiritual awareness. Yet there is a pedantry of the spirit in dwelling too much on these things. The force of new life can break through where and when we least expect it; as in Paris in May 1968, when the impossible seemed suddenly within reach.

It should at least be clear that to speak of 'good' design is to speak of, and from, the conditions of our own time, and our response to these conditions. The intelligibility – and perhaps the existence – of a design 'language' is a problem of the cultural fragmentation that affects participation in every other aspect of our culture. Because the realization of a designer's work is always socially contingent, his freedoms are always a recognition of necessity in a most explicit way. An elegant design solution is one that meets all the apparent conditions with a pleasing economy of means. A fruitful solution co-opts the conditions into a new integration of meaning, whereby what was 'apparent' is seen to have been insufficient. Such answers have questions in them.

6   Problems with method

The case for analytical technique – of the simplest order – rests on the premise that you should find out what you are doing, or what you should be doing, before considering how to do it; and that this degree of foresight is normal to the role and situation of a designer. Such a claim is not unreasonable. In medical terms, for instance, a prescription will be a very hit-and-miss affair unless there has been a diagnosis; and in converting a building, a drawn plan will be rather vague unless it has been converted from an accurate survey. In either case there are known procedures, and techniques, that will help to avoid wasted time and effort. So much is very nearly self-evident. Methodology goes further, by suggesting that preferred models are available for the definition of goals and the allocation of effort, and that if they are correctly employed, the designer's work can be guided towards an optimum solution that might otherwise have eluded him, except through the mediation of happy accident.

This is a somewhat crass simplification of the methodological position, and indeed to discuss it at all is to invite simplification; partly to keep clear of the dead language that the subject seems to attract, and partly to sidestep the quicksands of theory and debate that surround problem-solving and methodology, keeping most designers thoughtfully at a distance. This is a partly abandoned territory, and strewn with old workings – though there is life in some of them yet. Apart from the remaining devotees, who work out of sight and occasionally surface with new formulations, there are always small bands of students, picking over the discarded hopes and new offerings. The idea that the myth of creativity can be domesticated, and perhaps told how to behave – in terms at once 'scientific' and 'verifiable' – seems always to fascinate those who feel somehow denied by the myth, or outside its provenance. Jung speaks of the alchemists in this context, and the matter is discussed exhaustively (for ordinary mortals) in Arthur Koestler's *The act of creation,* a work that could usefully head the more specialized book references in the field, but rarely does. After all, Koestler wrote that fine novel *Darkness at noon,* which helps to argue for his credibility as a guide.

However, if there are real dangers in taking methodology and problem-solving too seriously, the supposed freedoms available from randomly 'intuitive' approaches are notoriously treacherous when the work is technically and functionally complex or demanding. Unless properly guided by diagnostic technique the outcome will tend to lack both objectivity and depth of thought, and in failing to break through invisible cordons of mental sets and prior assumptions, can become little more than a thinly rationalized ego-projection. It can also be said for methodology that in an often crooked way, its heart is in the right place. Even if a mouse-like issue is the result of seemingly elephantine labours, and even if in the process, a simple opportunity is mistaken for a 'problem', at least the effort is properly outward-looking; an effort of service. The other way of looking at it is to go back to part 1 and simply say that procedures appropriate to designing textiles will hardly do for the design of an operating theatre. Fortunately the breadth of the design spectrum is very accommodating, not only to a variety of opportunities, but to many different kinds of person. This part will inspect a few small-scale design situations to see how they look, and how they measure up; only thus will the relevant design attitudes fall into place, and seem less abstract in discussion.

Let us begin with chair design – although small in scale, one of the most difficult of design assignments. I remember discussing this with a Danish architect. He said that after ten years practice he felt he should be ready to try designing a chair. It was a failure. Now, five years later, he felt he was further along his road and perhaps it was time to have another go . . . Of course, if you are making something to please yourself, you can freely set your own terms of reference (as perhaps this designer was doing). This will not make things easier than working to a commission; merely different. You can decide, for instance, if it takes you that way, to make a chair the function of which is to question the notion of comfort, and which is therefore designedly uncomfortable to the point of actually collapsing when sat in. (A 'sign for sitting' with question mark added.) This may not be as daft as it sounds. Designers get accustomed to the support of many constraints (limiting conditions) in the way they normally work. Remove these and it is like putting an engine out of gear suddenly; it returns to idling speed, or feels purposeless when accelerated. One way to get the measure of a job, and to get close to it, is to project its life to absurdity (in all directions: scale, function, material, etc.) and then to pull back to some sense of boundary in what you propose to do. This technique is useful in many design situations. It was illustrated for me by a marvellous musician and a great teacher (Niso Ticciati) from whom I once had the privilege of some cello lessons. He said I should always try to form a mental picture of the whole arc of traverse of which the body, arm, and bow movement, were capable. When the bow then engaged a string briefly, for a short distance, to form a musical note – and given enough practice – the sound would come to be full and free and true. If on the other hand I approached with a picture or *Gestalt* of just the right size, and length of contact between bow and string, and therefore unconsciously aimed for that, the sound would be tight, thin, and restricted. He put it much better. The principle is worth dwelling upon in the imagination; it has many applications.

Suppose alternatively that a designer is commissioned to design (say) a polypropylene chair for factory production. The designer (you) will be involved in an intricate investigation of the possible – in terms of marketing, costs, production techniques, user requirements – that will compel you to conduct research in a fairly methodical way. You will be acting as an agent in a social transaction involving other people's lives and jobs and interests. It will certainly help to have guidelines for your own allocation of time and energy. The form of the chair – assuming 'normal' standards of comfort – will be to some extent preconditioned by what you find out about the human body, and where and how the chair might be used. Materials and form will be impinged upon by structural and other performance considerations, some of which may not be revealed except after the prolonged testing of a prototype, and also by the costs of tooling up for the job. The chair may be part of a 'set'; some grouping of furniture intended (now or in the future) to go together, in which case you will have to devise a formal vocabulary that will fit all cases (if loosely). The job would not be untypical of product design for a retail market. Such work involves a high degree of generality – the highest common denominators of everybody's needs, *as interpreted* by the buyers' estimate of market demand. (The 'buyer' is the person in the retail trade who places the orders.)

A simple problem analysis would help you with this job, and would net for you, for instance, the required life of the chair as a genuine condition to first question and subsequently work with. However, the scale and nature of the work would hardly demand more analysis than would come naturally as a refinement of common sense. Your real difficulty would be in satisfying the required conditions, having established what they are, and then designing what you felt to be a good chair out of those conditions. It could turn out that just one condition was paramount. To bring this factor into the light might well employ some of the techniques discussed further on in this book. More likely, however, 'good' for you would have to be defined as an educated designer's appreciation of chair-form nicely realized out of all the limiting conditions, one of which would be its acceptability to a less design-educated public. Tooling costs might be so heavy that you would have to be sure about this. On the other hand, you would not be asked to exercise unsupported judgement on issues outside your competence; if anything, your competence may be underestimated if you fail to ask searching and intelligent questions right at the outset. Here is a fairly tight situation for a designer; his margins of freedom are tightly drawn. All the more reason to find out how many of the apparent restrictions are real, inevitable; and how many merely reflect the habits and prejudices of your employer.

This could become acutely relevant in any design job outside the special conditions that attend product design. Building design, or say the interior design of some public space might offer you far more latitude for responsible decision. Suppose your client has asked for chairs but your analysis of the situation suggests that chairs are quite unnecessary – the needs of this area have been wrongly seen. Here it will be the seating *situation* that you examine. It is always a designer's job to hold in mind two counter-propositions: in this case, the particular notion of 'chair', as against the general notion of 'sitting' – only by departing from the familiar and particular case, can the designer freshly appraise its context and meaning; returning, perhaps, with a modified view of 'chair', or perhaps something quite different. New relationships may have emerged which alter the terms of the argument.

Drawing again upon personal experience, I was once asked to refurnish such a public area in a much used part of a college building. The area was noisy and unattractive and the existing furniture was badly knocked about. It soon became clear that if people were sitting down at all it was almost certainly in the wrong place and in the worst conditions for their comfort. By observation and questioning and taking opinions it began to emerge that this furnishing difficulty was merely symptomatic of problems that were located elsewhere. It was necessary not only to trace the circulation habits and patterns through this part of the building, but to ask more basic questions. In the end – and to cut a long story short – the whole area was replanned to include new student common rooms, reception rooms for guests, a public information area, a coffee-bar, and various other provisions. This in turn required the whole college to be reappraised in the light of long-term proposals for improving staff-student relationships so far as they were affected by physical environment. This required considerable alterations to offices, entrances, and other facilities. The proposals, and all the reasons supporting them, were sent out in report form to the various interests involved, and as may be imagined, much discussion took place before the *scale* of the problems, and their unfamiliarity, became generally recognized. In this case a problem analysis has begun with a small and apparently self-contained problem, but has finally so changed its terms of reference that 'chairs' have long since disappeared from detailed consideration (though at some point seating design will resurface as a practical design opportunity). Not only has the job become larger, but different in kind; questions of human needs and human behaviour have taken over from structural or material or aesthetic preoccupations (to return, of course, in new permutations).

Before leaving this example it is worth adding that if the client or the community as a whole had rejected the proposals, or been unable to afford them, it would still have been right to place any temporary action within a future perspective, and within a broad examination of the present. No use (on this argument) just treating a symptom. Yet there are limits to the 'broad view' – even cost limits – and there are also dangers attached to it. As with leaders or even dictators – one in ten may be benign, compassionate, and constructive, whereas the rest may be hell. The designer with the Napoleonic vision may actually be rather a bad designer, however correct his analysis. Here the attractions of a piecemeal solution become evident, and recall those of medicine as against surgery; as is said of the surgeons, they love their work too well. And surgery can go wrong . . .

There are, of course, alternatives. The designer might recommend a series of happenings or improvisations that would be in constant change; thus passing the buck to one form of participatory decision-making. He would not be designing any 'thing' as such, but recommending and arguing for the course of action that seemed most appropriate, and that would continue to pose the problem rather than answer it. Or he might suggest an interim solution; a set of more controlled experiments to see what can be done with the spaces, thus gradually accumulating evidence for a more far-reaching and a more structured solution at some time in the future. By this means the constants in the situation would become gradually manifest – constants, mostly, of behaviour and expectation – and the variables could be left to look after themselves. This is a bit like leaving a much-used swing door without a finger plate, in order to see what pattern of finger-marking builds up, before deciding what to do about it. A third alternative is quite common – to system-build from commercially available kits of parts, which can be dismantled and rebuilt or adjusted at will. This most reticent of alternatives (or cop-outs, as some designers might think) is seemingly in the customers' best interests. However, in practice, the most 'flexible' solutions, theoretically at the free disposal of the user, can come to feel either denuded of life, or psychologically tyrannous (or both); while an apparently rigid and very definite organization of a space can bestow freedom.

It is like living with all one's furniture (and maybe house as well) made out of perforated metal section; there is a continuous subliminal suggestion that you might be missing out on some better permutation of the parts. For qualities of *definiteness* in environments, it is worth examining some of the hill towns in Italy. The freedom bestowed by a walled town is – like some systems of belief – protectively finite; it is also that of recognition (by territorial boundaries, a most animal requirement). Presupposed, however, is a sense of necessity, of belonging, of history (therefore, of implied continuity into a future), and of an aggregate slowly cohering into what is finally experienced as an organic whole. It is a mistake to think that vernacular can be turned on at will, as a formal option; or made in any way easily at all, when in fact it is grown; but the study of vernacular form is more than a consolation of philosophy, it is a source of joy to the spirit and the senses. Our own situation is closer to 'seek simplicity and mistrust it'. A brand new office block is no match for such assets, however well designed. In part (but not, alas, in whole) this failing is a function of the papery and repetitious purposes that such buildings must serve – and thus, of a society that deploys its energies so barrenly. There are (incidentally) a few meliorist tactics for the softening of hard architecture, discussed by Robert Sommer in his book *Tight spaces.*

Turning to lighter stuff, there are some forms of display and exhibition design so volatile, colourful, and impermanent, that anything so ponderous as 'analysis' would strike them stone dead. This is the realm of theatre and the market place. An exhibition stand of this kind often succeeds or fails by the *élan vital* of a single overriding idea, to which everything else is strictly subordinate. The 'idea', however, will still have to be proposed, defended, found feasible, and – worst of all – realized. It is one thing to have a brilliant idea, quite another to make it work. Thus even here, a designer will find that he is not exempt from a good deal of hard thinking and from all the communication and other techniques that forward, prompt, order, and regulate, such thinking. As is well known (or should be), if an argument persuades by its simplicity, usually someone has done their homework: the most economical design may conceal the most arduous preparation. Other and more sober forms of exhibition design (museum display is an example) are normally quite outside the possibility of improvisation, and need a most carefully balanced appraisal of every factor relevant to the design. Exhibitions with a mainly informative content may sometimes go to extreme limits just to avoid being dull, and therefore non-communicative. Such weighting is well within the middle-spectrum range of ordinary design problems. It remains true, however, that exhibition design generally is wide-latitude design; often enough it gets away with little more than a well-acted conviction of purpose.

Graphic designers very often begin with a false brief from their client which will need to be taken right back to its origins before design (in the ordinary sense) can begin. This applies less to bookwork than to the very wide field in which designers are now employed. Exhibition design is an example. The graphic designer may have full responsibility, employing another designer with structural competence as an assistant or collaborator; or the roles may be reversed. Either way a root concept will have to be mutually understood and discussed, and the design may depart far from an initial briefing. Often there is a vagueness of function in exhibition design – the exhibitor may just want to 'be seen to be present' but in a favourable light; but may not put it (or see it) quite in that way. Whatever the requirements turn out to be, it is clear that the specialists involved must understand each other, and to do that they must speak the same language, at least up to the point where technical expertise begins to take over. Such examples point to the dubious logic of educating designers in compartments, and of over-using words like 'two-dimensional'. When was an exhibition ever a two-dimensional experience, or for that matter, catalogues, and other forms of jobbing graphics?

When design decisions are very open and unconstrained – and this might occur in any field of design, not merely in ceramics and textiles where it is usually assumed to occur – it is all the more necessary that the designer has some shared sense of context with other specialists. This need not imply one way or one culture, but does imply a history and a pedigree in any cultural effort. Nobody starts from nowhere; in which case it is better to determine your sources of inspiration, to go after them, and to try to understand what they are about. Otherwise a designer's vision is simply being programmed by 'the surface of the face of things'. It would be silly, for instance, to start designing modern fabrics intended for modern buildings with no understanding whatever of modern architecture. A true understanding involves more than a few images of interiors. It involves what has been attempted in the whole field of object-relations, and the questions, hopes, or assumptions that underpin modern design – that make it intelligible. The textile designer (or other specialist) who conceives the work as self-expression, and works solely with other like-minded specialists, may make or design some very nice things. In the long run, however, the work will be deprived of connection, of 'due weight' in its context of use, of appropriateness; it will tend to become, in short, self-contained. The opposite was true of some of the very subtle block-stripe curtain materials that formerly came from Scandinavia; such material reaches out to its possible context and does not feel 'complete' without it – a property, partly, of directional flow but not peculiar to linear design.

Designing for a missing factor – that of use – is an interesting matter. It is continuously a problem in chair design, where too anatomically receptive a form will seem intrusively empty when nobody is sitting in it – which may mean for most of its life. Its immediate neighbours will be reinforcing the problem. This is one reason why Thonet chairs and stickbacks work so well spatially; they seem to comb space with just a sufficiently forward reminder of their existence. Space flows through them.

A much smaller and formally perfect example is the old Pelikan Graphos pen (as one now has to say, nib-holder) – always a joy to hold and to *be*hold. This pen is a cylinder so pure that aficionados find ample variants in the doming of the ends (the radius of which can provoke discussion), the length of the pocket clip, which unaccountably varies, and the barely-visible placing of the maker's imprint. The clip is itself a *tour de force* executed without break in the formal language. Here is fine-tolerance decision-making, though coarse by engineering standards, and it is against the severity of the cylinder that the user notes an appropriate weakening of one end, where a screw thread for the cap is turned. Thus only when the pen is used, and the cap duly screwed in place, is the integrity of the cylinder fully asserted. However the geometry is of course weakened at the business end where the nib is now revealed (nibs are individually inserted): the joy, of course, is the transference from formal language to active function, so that what might seem visually weak is suddenly felt as intelligently strong; a case of function compensating form. The fact that the pen works so well (though it is not easy to use, the two are not the same), in that it will produce with precision lines from $\frac{1}{4}''$ thick to a hairline in equally precise gradations, provides a functional authority against which all these small points gain in the telling. However, the firm was taken over, the new version works just as well, employs the same principle, and has been 'restyled'. I say no more.

Here then are considerations of some subtlety whereby one design is found acceptable as against its functionally identical counterpart. In designing the pen in the first place, what 'procedures' or 'methods' would bring a designer towards such a distinction, help him to perceive it, and encourage the perception to become deeply felt? This is one thing to think about; the other is the scale and nature of the distinctions, and the power in them, *once perceived*.

In the general nature of formal disciplines, it is worth mentioning a common and vulgar heresy rife in some of the colleges, and sometimes supported with more vigour than understanding by Assessors, Course Validators, etc: it is to suppose that every design decision must of its nature demonstrate 'fresh and original thinking', and that the work must be coarse-grained – bear the marks of travail over a wide area of decision-making. This view is partly an unhappy legacy of the competitive spirit that we take for granted in secondary schooling, and which goes on to bedevil higher education in almost all its manifestations. Such thinking has much to do with the serious mismatch between education and 'real life' whenever the gladiatorial degree-taking image is confused with the conditions of design practice. However, these matters are less relevant here than one practical consequence, which is a confusion of gross with fine decision-making. The distinction may in fact be one of gross and subtle judgement.

To illustrate this the Pelikan must be exchanged for a bicycle. As will be seen, the two objects have something in common, at least in the context of this discussion. Consider the situation of a student working on the design of a bicycle. Now this machine is a beautifully tested and developed type-form over a given range of performance conditions; it is known to be a remarkably efficient energy-converter, and it is indubitably the product of a vast amount of co-operative human endeavour to get it to its present stage of refinement. Is there any good reason to redesign the bicycle from first principles? The answer must surely be – probably not, unless you posit quite new conditions of use to which the standard design is ill-adapted. It might be argued that there are such conditions, such as the effects of parking and fuel problems on commuter traffic. Designers have addressed themselves once again to the folding portable bicycle, which commuters and others might take on public transport. The folding Bickerton with its carrying bag is one light-weight answer; a bit fiddly and doubtless the principle is capable of further development; other and more seemingly radical designs have been published, but not it seems yet marketed. The Bickerton uses conventional spoke wheels and other standard components; it is to that extent a compromise, perhaps, but it works well. There is an inevitable loss of efficiency once the machine is rolling, as compared with a full-size light-weight cycle; this has always been the case with small-wheeled machines and for good reasons. No doubt there will be further work on special purpose bicycles where rolling efficiency can be set against more compelling criteria.

However, the fact remains that *within* a developed type-form there is still a very great subtlety of possible variants and a considerable range for them. In the case of the bicycle, this applies not only to matters like the design of accessories to develop possible uses in yet unsuspected directions, but also to less obvious matters. The design of the front forks, for example: to get the curve just right, not only by gross functional criteria but also by eye, is no trivial matter. A comparison could be made with cylindrical pots of identical form on which the glaze is progressively varied, or which are very slightly modified in their proportions. Such adjustments are of the essence in realizing a design far beyond the diagram of its intended life (which can be intellectually appraised) and towards its final form; to experience such work can be deeply educative. Considerations of texture, scale, colour, tone, and the radius of an edge; the quality of a curve known so well to a boatbuilder in fairing off a sheer-strake: in getting these things right, or as right as judgement has to concede, the designer employs a concentration, and an artistry, of a different order to that which earlier stages employ – with the aid of logic and rational enquiry. Those stages always seem incredibly far distant . . .

The designer who does not come to feel naturally for these things (given enough self-training) is as deprived as someone who says 'I don't fancy the physical side of my love-life'. There are many aids to such self-training. No opportunity should be lost for making comparisons. For instance, students should always look at walls attentively, the plainer the better. Observe the fleeting differences of light and shade, surface, colour, and texture. Imagine one quality of surface replaced by another, and the difference it would make. See how a volume is affected by its planes and their surface qualities. Study brickwork and the effects not only of colour and texture gradations but of different types of pointing. Become aware that within the possibilities of raked pointing there are many distinctions. Confront some of the best small Georgian brick houses; see them first through half-closed eyes just as boxes with perforated holes; then gradually allow the subtlety of detailing, material, and proportion, to establish itself. In typography such matters are taken for granted. Letter forms are finely distinguished and letter-spacing of capitals is still practised, though by no means invariably. In asymmetric typography the possibilities are clearer, because logical distinctions and heirarchies of meaning can be pointed up by very small adjustments of relationship on the page. In raising consciousness of these matters, it should be remembered that our civilization *sells itself through sensation,* preferably with the volume turned up. This is good reason for designers to learn how to speak quietly, and to understand how it is that conversation becomes possible between people and things. There are good illustrated books to help with such studies. Examples are Gordon Cullen's *Townscape,* Elisabeth Beazley's *Design and detail of the space between buildings,* and Peter Smithson's *Bath: walks within walls.*

Returning to another false definition of design capability, it is surely foolish to suppose that everything in the world must be continuously remade or rebuilt, yet students are made to feel inadequate if they do anything less. Such pretensions must involve quite a waste of human energy and much disappointment; they also demean the possibilities we have for learning from each other. It is becoming recognized, slowly, that buildings should be readapted to fresh uses wherever possible; the poor have always involuntarily recycled kids' toys and clothing. There are still many appliances that could do with improving, bringing up to date, or simply resuscitating. This has happened recently to some extent with wood-burning and other solid-fuel stoves; in the same area, there are many efficient paraffin devices awaiting rescue. The requirement may sometimes seem to be that of 'styling', when (say) a given mechanical principle is accepted without serious change. In a sense it is, but not in the way of imposing a style, or a concealment, upon a naked mechanism, as finding the clothes that may suit, and express, its own internal life. If this is a little too anthropomorphic, the fault is here on the right side. This is only one of many possible design activities which (like interior design) are interpretative in their nature, and which could offer a contribution within the spectrum of 'adapt, adjust, make do, mend, and develop': as distinct from 'erase, forget, and innovate'. Much can be learned from this humane preoccupation. As for colleges that stress innovation and 'originality' far too heavily, it is doubtful whether the 'look Ma no hands' content of a degree show will ever lose its appeal (to students, their relatives, and college staff), but perhaps these occasions might be put on a more kindly footing. Before leaving the subject, it might be said – and not *too* quietly – that the modern movement has always opposed the pettiness and irrelevance of competitive originality, and can well support the propositions in these paragraphs without historical inconsistency; though, if this were not the case, it would also be within the spirit of the modern movement to change its mind . . .

Two further examples must complete the account of design possibilities begun in this part. A furniture designer is considering the design of a sideboard, and let it be supposed, to simplify matters, that the design brief is very open. If you were the designer, how might you view such an opportunity? To start with you would probably have ideas of your own, but let these be put to one side (not to be lost, but returned to). You will know that the word 'sideboard' means that a certain commonly recognizable piece of furniture has emerged out of history. The origins of the piece are interesting, and perhaps suggestive, but you will question whether these old uses now exist, or exist in that form. Do the furnishing contexts of houses today, and the way people now live, suggest that 'sideboards' still do the same job? You might recall Rietveld's work on this problem, and what it had to say so presciently: 'here is a contingent equation, the elements and components of a sideboard reaching out and dissolving away into new meanings, new relationships with what surrounds them, a question implicit in an answer' (I hasten to say that this is my own response, not Rietveld's). What will be the surrounding conditions? As has been said about chair design, in practice the brief will close around the limits of the possible: marketing policy, a 'range of furniture', estimated demand, and so forth, are the governing factors in the retail market. User-context does not exist, is unknowable, except indirectly through the operation of supply and demand: furniture is really designed for the shop buyers who make sales policy. It is inevitable that design takes place on a stylistic footing, ringing the changes annually to stimulate the market. The possibilities widen in some of the more specialized or design-conscious shops (for example, in London, Heal's and Habitat), and of course dramatically so in the one-off sector – which is where it all happens for design that is formally innovative.

It should be clear that beyond the doodles in a designer's sketchbook there is no such thing as an open brief. There will always be limitations of context if not of contract; and those of cost, materials, and the special conditions of manufacture. The identification of such limits, and the turning of them to constructive account, goes hand in hand with an analysis of use, of performance. In the case of the sideboard, just what is a 'board' and a 'side'? If it is a storage cabinet, what is being stored, why, for whom, for what purposes? Are the contents used regularly – all the contents, or are some virtually dead storage? Should they be there at all? What order and kind of accessibility is implied? Why are the contents screened? – to keep out dust, or to keep out of sight and mind? Is there a temperature or humidity problem? Might there be? What is the 'top' and what is it used for? Is there necessarily a back and a front? In answering these questions and many more, it may be helpful to use a degree of abstraction as a tool of analysis – or perhaps it is better to say, as an analytical resource. For instance, for my own work I devised the categories of surface, container, support, threshold, screen, connector, traverse, grip, and mechanism. By such analytical procedures it is possible to dismantle an object which may have lost its reasons for existence, and to reconstruct it in new terms. The object as such may disappear if its meaning is better seen in some more selective or more comprehensive arrangement of parts – 'parts' being here, the purposes served. Thus, for instance, wardrobes disappear into storage walls – conceptually, but not on the retail market, which remains object-fixated although tending to respond to appearances.

For designers the real difficulties do not occur at an analytical level – the determining of relationships – but in the realization of a design, against criteria that become increasingly elusive. This is the realm of the senses, of imagination, and of judgement. The design *may* exhibit an intellectual proposition, or an argument that can be 'read' and for which the key premises may be available to inference. This is unlikely to be more than one dimension of meaning in the job and rarely an essential one – except to the designer, whose satisfaction it may be to write in to the job a clear level of argument as an optional reading, free of any impairment to its more ordinary usefulness. In his analytical work the designer seeks a sense of priority in the work and a full understanding of the conditions that must determine its life. He may seek also a 'diagram of correspondence' to help him – not to dodge his constraints, to get round them, but to take them up, to identify their potential, and to use this potential to positive advantage. As has been said earlier, this conversion of constraint into opportunity is about as good a definition of design as any. The sideboard may well disappear as a recognizable entity and its functions disappear into new relationships. It should be noted, however, how many presuppositions of economic openness have been necessary even to make this discussion possible.

For social and economic reasons design is far from being an open-ended activity; yet, it is an insistent requirement of modern design that openness and intelligibility should characterize its products; that life should be free to flow through them, that their meaning should be defined – indeed, made accessible, by projections of reference that are strictly outside the problem itself. Designers in some fields will have seen an object developing into a system, system into process, and process into information, in a way that disconcertingly suggests that their world is dissolving away around them. In making finite decisions they will be aware of a new dimension of reference in their thinking. This is the simplest consequence of a changing world-picture, expressed in the physical sciences, philosophy, the arts, communcations technology, and elsewhere. A highly particularized design solution – say a one-off job with clear determinants from place and occasion – must now not only exhibit a field of generic reference, but *derive a gain in particularity from it*. Once this design requirement is identified, felt for (it does not have to be an intellectual quest), it will become clear that the survival of modern design, and its character, is inseparable from the interdependence of modern man, the principles of mutuality that knit together his societies, and the altered view he now has of his world, as distinct from that of his village. He does indeed occupy perhaps more consciously a village, but with an altered frame of reference.

Such perceptions have their equivalent in design education. The textile designer who has never felt for the distinctions of outlook implied by (say) Ronchamp, the Barcelona Pavilion, and the Dymaxion house (a specially indigestible mix), or has never distinguished between the geometry of a cube and an inflated structure, is simply working as a badly educated specialist. If on the other hand, such understanding is achieved, there is no loss of focus or immediacy in the designer's own work, but it may well be helped to *refer outwards* and gain some reinforcement of meaning thereby. The designer's intuitions will be instructed by a richer sense of the place he occupies. There is a parallel with analytical procedure. In forming a hypothesis, the designer will not only have a picture of prior demand to test it against – this is professional necessity – but he will also find that when a hypothesis suddenly springs into consciousness, it is instructed, fitting the conditions of the job rather than arbitrary conditions – and this is creative necessity. When something goes wrong, it can usually be traced back to the beginning, from the acceptance of false premises. Hence on the one hand the importance of questions, and on the other, of the resourcefulness of attitude that prompts them. Such basic requirements (asking the right questions, knowing how to, and mustering the will) are far from being naturally coterminous. This is why technique must be kind to the user and not only 'technically effective'. To use a designer's figure of speech, if the cross-sectional area of personality fails to be exposed to the creative possibilities in a job, so that when necessary the designer can get his weight behind his imagination, then the situation can easily become a sterile one. Technique has a core-structure that is irreducible, and possessed by an elementary logic and sequence, but it can be used, and interpreted, with a great deal of personal latitude.

The last example of design opportunities is well discussed by Papanek and others (so it will be only touched upon briefly here), namely, that of inventive design in so many obvious areas of social and personal deprivation, where the look of something and the pleasure of handling or contemplating its qualities are still very relevant, but may have to be strictly secondary to producing it cheaply, getting it marketed, and ensuring that it works for its living. Equipment for the handicapped is a case in point. This work is often close to the engineering end of the design spectrum. Advocates of such work sometimes seem to twist the arm of students into thinking that *only* preoccupations of this kind are socially responsible. This is a false view, though not unsympathetic. It might be said that two design jobs in the first half of this century have been brilliantly apt, inventive, and economical. One is the design of the cat's-eye for road transport and the other is the design of the London Underground map. In fact the cat's-eye considered merely as an item of manufacture, is a very nice piece of anonymous sculpture, and its self-cleaning rubber eye-lid is a concept masterly in its essential simplicity – but nobody digs up cat's-eyes to admire their form, the important thing is that they work so extraordinarily well. The one job is superbly inventive, the other a first-class ordering of information – a model of its kind. However, it is a mistake to suppose that designers can freely elect to design the equivalents for cat's-eyes, and London Underground maps, or mechanical equipment for handicapped children. The same goes for the trailing link baby-buggy, which must have eased life's burden for so many harassed mothers getting on and off buses with children, or out shopping. An aptitude for mechanical invention is one kind of design skill and it is also a personal design aptitude; not all designers *can* work in this way, or are naturally drawn to it, whatever view they may take of the moral imperatives that seem to beckon their commitment. Designers must find out just where, within the very wide design spectrum, their own individual contribution can be most effective.

Several conclusions may be drawn – tentatively – from this discussion. The first is that design opportunities are so various and plentiful, and designers so different in their natural capability, that it is rash to generalize too far about the way in which they do (or should) approach their work. Second, it is clear that analytical thinking may profit from insights that are interdisciplinary; practical design procedures are necessarily so. This is certainly the case up to that familiar and pleasing stage in design work, when the categories of requirement in a job suddenly seem to be matching nicely with the categories of a possible answer. The third conclusion is that this kind of thinking, together with simple problem-solving *procedure*, is quite natural to design work, but not rigidly or predictably so, over the middle range of the design spectrum (see part 1) – and hardly at all in certain areas. The productive relevance of any design procedure is partly a function of the designer's personality; procedures only exist for people who employ them. Other things being equal, those most acceptable will be closest to the exercise of judgement in ordinary life. If this is so, then it may well be that the ordinary syntactic 'thinking-out-loud' and relatively loose-fit and two-way devices like the report (see part 17) have been somewhat underestimated in favour of logical models of greater sophistication, even though such models have had to be partly dismantled to accommodate objections. In short, it is probably useful to recognize a distinction between technique and method, taking the latter no further than the more lenient notion of 'procedure'.

The natural limitation of such a view is the range of work with which this book is properly concerned. Within this range it is always tempting to confuse logical modes of thinking with the untidy and informal procedures that usually prompt fruitful ideas into being. As is commonly known, the best ideas often occur at odd moments on the backs of old envelopes. Computer-aided analytical technique becomes necessary as the scale of the work increases – and moves off the stage of this book – and as the conditions for its measurable efficiency become both critical and manifold. Where very complex constraints are concerned, as in planning the co-ordination of operations on a large building site, the job cannot be fumbled as far as an analysis of quantities is concerned. For any work with expressive latitude, it is still necessary to define its extent. But in the work discussed here, analytical thinking has the secondary (but crucial) function of presenting the designer to the work in a responsive frame of mind, and in good heart. For some people, reliance on sequential method will paralyse rather than quicken their responses.

The worst error is to take refuge in notions of 'method' or 'process' at the cost of any practical commitment. Thus is a wilderness created by default, and argued for in retrospect by a specious appeal to scientific method in the way the problem was approached. Sir Peter Medawar has reminded us that the scientific hypothesis takes the form of a hunch, or an intelligent guess, which is *then* scrupulously examined against existing schemata to see if it works. Not that such comparisons throw a direct light upon the designer's role, which is to be available to tasks for which his experience is fitting, in full awareness that some things are much better done by people for themselves.

A socially adaptive view of design practice takes the view that a designer accepts what he is given, and makes the best of it, by knowing how to weigh up his constraints shrewdly and logically. This is a perilously inadequate view of adaptation, as can be seen in every area of society where new adaptations to reality are breaking through old and outworn assumptions. Designers must themselves adapt to a changing picture of their professional role and its validity. As was argued on different grounds in part 5, the modern movement advances by fruitful hypothesis, which is often code-breaking. It steps sideways, by an elegant or 'fitting' alternative; and it steps backward by keeping up appearances. Fortunately, falling into disrepute a little, its adherents can relax and accept their common human failings.

Choosing, therefore, a more homely language: a designer must have his tools around him in the ordinary way of competence. A good joiner, who knows (as the saying is) seasoned wood from green, will have in his toolkit a few tools that are rarely used but irreplaceable when needed; he must know how to use them and when to pick them up. A designer's priority is openness of creative response, the capacity to mix the levels of his thinking, to ask productive questions; he must seek the personal conditions that help him to work in that way, and the social conditions that allow him to. This search will not prevent him getting on with the job at the limits of the possible, keeping all his tools sharp for the next chance to make himself useful.

The artist

The artist: disciple, abundant, multiple, restless.
The true artist: capable, practising, skilful;
maintains dialogue with his heart, meets things with his mind.

The true artist: draws out all from his heart,
works with delight, makes things with calm, with sagacity,
works like a true Toltec, composes his objects, works dexterously,
arranges materials, adorns them, makes them adjust, invents.

The carrion artist: works at random, sneers at the people,
makes things opaque, brushes across the surface of the face of things,
works without care, defrauds people, is a thief.

Toltec poem translated from the Spanish by Denise Levertov

7   Designer as artisan

It is well known that many designers have their own small workshop, for modelmaking, for making up and testing full-size details, for prototype work, and sometimes simply for play – which is not only a relief from the drawing board, but can prove directly and unexpectedly helpful to any job on hand. Industrial product designers will usually have access to factory facilities for prototype work, or may co-operate with the research and development group already attached to the factory. In Scandinavia, it is not uncommon to find designers operating their own studio/workshops in direct association with a manufacturer – a large proportion of their time being given to the production needs of the factory, but some to their own more personal work, which may, or may not, ultimately serve the interests of their hosts. This degree of subsidized independence is a shrewdly practical arrangement on both sides, because the manufacturer will pick up any benefit from a designer whose work is going well; but it does imply a restricted field, and ties that a designer may not always find acceptable. It works best for specialists (in glass, pottery, furniture, for example). Obviously a product designer will always need close contact with a client's technical staff, if he is working to a tight brief and to all the limiting conditions of a particular production set-up and its marketing potential. In these circumstances the private workshop will be less useful – or even misleading. However, it is a modest technical base within which to experiment and improvise.

The artisan designer is a different animal altogether, often equally remote from the drawing board and from the corridors of management; engaging directly in the business of small-scale production which may necessarily include marketing and distribution. The artisan is a blue-collar worker; he wears overalls (or needs to), has dust in his hair, and gets his hands dirty. Inescapably, he partakes of behaviour patterns that distinguish working men from higher management, though he will earn rather less, work longer hours, and will approach his work with aims and standards that tradesmen will not normally articulate, or even find sympathetic. Thus among fellow workers the artisan will feel somewhat estranged, though sharing, at some levels, much of their culture; with business men and white-collar designers (particularly architects) he may feel equally uncomfortable, though for different reasons. For academic and design theorists he will feel the least sympathy – although artisans often teach – probably because they seem to him to speak from a distasteful blend of economic privilege and practical ignorance. Well-off left-wingers are unpopular for similar reasons. The artisan identifies strongly with the 'shop floor'; anyone who pontificates without having done their time at the bench, or minding a machine, is felt to be automatically disqualified, and not to be taken seriously.

Thus it will be seen that the working stance of an artisan can become at once isolated and prone to internal contradictions, for reasons that are inherently class-related. Add to this his continuously insecure economic footing, and the contradictions implicit in his marketing problems (at worst, he may be producing peasant-revival knick-knacks for middle-class weekenders), and it is then possible to see how so many small workshops are cut off from the intellectual currents of their time, and produce work that is self-contained, sentimental, and backward-looking. However, in the late 1960s and in the 1970s there was a revival of enthusiasm for this way of working. This followed naturally and properly enough from the energy crisis (or rather, the abrupt public awakening to its existence), a renewed interest in conservation and the need for a decentralized economy, and from the growth of attitudes celebrated, typically, in E.F.Schumacher's *Small is beautiful,* and in the writings of Ivan Illich. That all this was placed in a forward-looking context by Kropotkin (*Fields, factories and workshops*) writing in 1898, is neither here nor there. More interesting for present purposes is the coming-full-circle of the modern movement, and what seems to be the historical moment for a reappraisal of this way of working. Anyone who doubts the valid *scale* of small workshops (assuming they survive at all) should not only contemplate the miniscule attainment of modern design through factory production over the past fifty years, but should read Percy Mitchell's *A boatbuilder's story* – a real eye-opener as to what can be done by a few men working with primitive equipment and against unbelievable practical hardships. These are the extremes: at a less strenuous level, and given that an artisan's life is problematical, over-worked, and under-paid, what reasons might there be for a designer choosing it?

Elective affinity is the most understandable reason – a liking for tools and materials and for the *grounded* nature of this way of working;

with its converse, a disinclination to work so remotely from the
final product, a dislike of paper work, and a mistrust of offices and all
that they can seem to imply of the clerkly life. Against the attractions
of workshops, which are only too easy to romanticize, must
be set the severe limitations of scale that any workshop must accept
(and which some designers find frustrating), the long hours, and all
the anxieties and privations that attend self-employment at a relatively
low standard of living. Such constraints will prove a stimulus to some
people and a discouragement to others. The skills required are –
surprisingly – not much of a problem. Trade practice is always
hedged about with thorny growths of technical mystification, so that
'their' patch is not too easily trodden over or under-valued. In college
technical shops, the women starting from scratch are always surprised
by how quickly they come to feel at home. This is not to say that
standards are easy to establish and maintain, or that craft skills
demand much less than a working lifetime for their full mastery; but
an intelligent and willing person who really concentrates, can learn
the elements of workshop practice in far less time than is taken for a
full trade apprenticeship (granted, admittedly, some initial aptitude,
which it is fair to presuppose).

Some designers are driven to experiment with workshops, often quite
late in their working life, in an emotional response to the logic of
their own perceptions. They may feel that in western societies, the
only people who can afford design services are those who already
have more than they need, and that therefore, 'professional' designing
is a form of cake decoration when the rest of the world needs bread.
Alternatively, and relatedly, they may decide that managerial
posturing is no way forward in the frequent confrontations between
workers and management, and that the decent thing is to withdraw
from the prior status-assurance (and salary structure) that the
professions are in being to ensure for their members. It must be said
at once that artisans have no way of opting out of their society –
except into the illusion of total self-sufficiency, that final parody of
the mutual aid principle – because the alternative networks of
distribution and exchange do not in fact exist (or not as yet) except in
small and occasional areas of social experimentation. Workshops
depend like everybody else on an open-market economy.

However, there are certain existential advantages in the workshop
situation. First, a service workshop (see below) can integrate itself
into a local community in a very direct way. Such a workshop grows

and proceeds through personal encounters; which keeps every option open and, in a sense, renewed. The opportunities for experimental design work will be, proportionately, few; but their context will be organically provided for in a possible widening of acceptance as the workshop gradually builds up local confidence in its services. Second, in any kind of workshop there is a natural give and take that acts as a control, giving immediacy and proportion to its activities; and there is continuous feedback both in the way of working, and probably through the marketing arrangements. Third, in the cultural desert that we would all like to fertilize with new vernacular life, it may be more satisfactory to make things (and to make them well) than to talk about it, or to issue instructions.

Broadly speaking there are two ways of working as an artisan: with a service workshop, or in production. In the former case the workshop serves local needs and is open to whatever comes through the door on Monday mornings – just as a garage is. The service workshop emphasizes the one-off product (in the case of a joinery shop this might be anything from a staircase to a shop-fitting job or an exhibition stand) and occasionally the service as such (e.g. repairs and renewals); there may also be a production bread-and-butter line to help keep things going. The production shop is far more common and of course for many years studio potters have worked in this way, which (by simple definition) implies that the designer has a range of products in mind to begin with, or that possibility as an aim, and then sets up a workshop to make them. Other things being equal, it is very desirable that the production workshop has its own retail shop. Transport costs are reduced (the customers do the travelling), middle-men's (legitimate) profits are eliminated, packing and storage problems are minimized, other people's products can be sold, and that mysterious entity 'goodwill' is fostered, because your customers come back to you (products sold on the open market simply disappear). Shops are common as boutiques, or attached to potteries, but less common elsewhere (perhaps because other things rarely *are* equal).

It seems to be the general experience that there is no lack of work for artisans in any field: the problem lies in sustaining the first two years. It is decidedly a mistake for design theorists to suppose that every serious design opportunity must refer back either to the type-form (a myth with Platonic overtones) or to the special conditions of mass production, when in fact other demands are real enough and can often accommodate a more flexible response, both as to design and to the means of production. Not only is 'custom' work (one-offs) an economic proposition, but very often short production runs are better handled by small producers. Such work will include jobbing printing, exhibition and shopfitting work, purpose-made furniture for hostels, schools, laboratories, hotels, etc, interior conversion work, package furniture, and other possibilities too numerous to mention. Production facilities will normally include as much mechanization as the shop can reasonably afford, but the machines will be of the open-ended type that extend the hand rather than replace it. (For example: a joinery shop of this sort would have a router, spindle, dovetailer, mortiser, in addition to the usual sawbench and planers etc; but would be unlikely to have a multiple jig borer, variable offset lathes, or a four-cutter.) It is a serious mistake to begin with portable electric handtools on the supposition that these machines will fulfil anything like their theoretical capabilities (and be otherwise attractive for their sheer portability): far better to spend the money on heavy second-hand equipment. This is almost a general law for all trades, and all knowledgeable tradesmen confirm it. Heavy castings soak up vibration and are safer and far less tiring to use: where cutters are involved, they will tend to hold their edge better where chatter is eliminated. However, such matters belong to the delights of homespun technical controversy and are outside the scope of this book.

What are the worst difficulties? The most serious is the need for adequate working capital. Long term loans on low-or-no interest terms may be available from CoSIRA (the Council for Small Industries in Rural Areas). This admirable body, formerly the Rural Industries Bureau, provides a comprehensive advisory service on all matters of interest to small workshops – from costing to machine maintenance – and through their local advisory officers, they are an excellent grapevine for valuable marketing and technical information. There are also grants or loans obtainable under schemes for artist craftsmen. (See part 18, 'Finding out for yourself'.)

Without some years of practical experience (meaning, doing it the hard way) it is common to take rather too rosy a view of the financial difficulties. Beginners overlook the need for reserves against slow payment (some reputable big stores take 3 to 6 months to pay), against bad debts (clients who may not pay at all) and the costs of advertising, transport, cleaning, postage, travel, etc, apart from the more obvious overheads like rent, heating, light, and power, the need to allow for holidays and sick time, the cost of time spent chatting to people and doing the book-keeping, and the many other matters that may tip the balance between ruin and success. Form-fillers and officials of all kinds are the nightmare anxieties of any artisan's life – as indeed of any self-employed person – and records tend to become scanty, episodic, or lost in brown paper bags under benches. For all these reasons it is wise to take advice early on from CoSIRA officers, who, like the best doctors, manage to combine impartiality with the impression of rooting for your own workshop against all comers.

The problem of locale may be set up, or solved, by personal considerations, or by happy accident; otherwise, and given an entirely open situation, it is more difficult. Workshops must keep in touch with their market, and enough customers, yet city premises have high overheads; a cheaper place in the country may either involve high transport costs, or may force the decision to service local demand. Beginners usually underestimate floor space. The problem here is keeping various operations and storage provisions sufficiently distinct, or actually sealed from each other in the case of dusty operations like wood machining. In a joinery shop for instance, assembly must be separated from machining and separate space may be needed both as a dust-free polishing and/or spray shop and perhaps for packing and storage. The latter will comprise both finished work and component parts. All this will naturally follow from the nature of the shop and its products, but this may not be easily predictable at the time that premises are secured. It is therefore wise to allow generously at the outset or to try to find premises with some expansion possibility. In the cities, co-operatives are possible; and of course workshops can be run in this way – indeed there is no limit to the variety of options available. Some of the craftier workshops – those that look back to the Gimson and Barnsley tradition of fine one-off pieces – have developed to its logical extreme the possible expensiveness of hand-built work (a deeply unattractive option but showing the range of the spectrum). Such a heavily stressed craft approach is no doubt a good survival strategy in its own terms. Some artisans can get a little part-time teaching to help out. Often this doesn't work, the two worlds being so far apart; but sometimes it does.

Another difficulty is that of work coming in at the wrong time. Turning down work makes for bad relations, delivering it very late is considerably worse, and rushing the job at cost to technical standards is worst of all. Under these circumstances, it is hard to find a balance between a properly equipped and economically viable small workshop, and its necessary escalation into small-scale factory production, with all the economic and personal pressures that this may imply. Some factories have begun in just this way – Conran's was an example – and have continued to maintain conscientious design standards, if at some cost to the original intentions of the designer; but at some point an artisan must decide in principle how much he wants to expand (if the possibility presents itself) and how such expansion might affect the personal goals he has set himself.

Having reviewed the possibilities, and the hazards attached to them, it remains to point out that very few workshops are informed by the kind of design consciousness that designers (as met with in this book) might promptly recognize. In Britain the prospect is said to have brightened during the 1970s. At the time of my own workshop (with George Philip, 1949 to 1959) — when such undertakings were considered treasonable to the modern movement — we were unable to find any workshop in Britain to swop notes with, though plenty doing their thing in the language of the Crafts Centre. Communication at that level was confined to technical matters (most craft workshops of that time worked to high technical standards, particularly in chairmaking and cabinetmaking). An exception to be celebrated is the printer Desmond Jeffrey, who did some good work without making too much fuss about it; and it is not well known that the influential typographer and teacher Anthony Froshaug did some of his best work on a workshop footing. This raises the fact that some designers will find it relevant to work in this way during some part of their working lives – perhaps for some of the reasons already suggested, but also because the problems in design that most interest them are best pursued by these means (which include, of course, the benefit of well understood production constraints). Rather different is the case of the Leach Pottery, which produced a standard range of domestic ware alongside the more individual and personally expressive pots by Bernard Leach; how far this arrangement formed an economic subsidy I do not know. David Mellor's cutlery workshop (or manufactory as he might prefer to call it), and his retail shops, are a straightforward instance of the 'production' alternative. Different again is the situation of designers who are able to use some of the resources to be found in college workshops; though here not only is the situation academically subsidized, but in the nature of things, direct production can only be occasional. There is of course the academic tradition of the limited edition, often of parochial interest to the college concerned, but when work of a wider appeal emerges from the presses its graphic design is rarely distinctive. The college prospectus is sometimes better conceived that its art printing. There are adventurous exceptions; an English example is seen in the work of the graphic designer Ken Campbell, who has printed his own books in London and Corsham.

This is the 'workshop situation' at its most attenuated – and, quite characteristically, at its closest both to academic life and to the possibility of subsidy, neither of which typify its ordinary economic independence. It is the day-to-day survival problem that gives the small production unit a certain realism of outlook, and helps to offset, perhaps, too much preciousness in its design approach. It will certainly be urged, however, that even at their most robust, the small workshops make too personalized a contribution to be of much general interest, and that their production potential is so small, that designers will hardly wish to take them seriously. To this charge, an artisan might reply 'never mind the width, mate, feel the quality' – and then, looking more closely, he might add: 'where *is* this width anyway; what can you show for your vastly superior resources?' At a deeper level of argument, the notional value of direct production is defensible on rather different grounds, with less emphasis on the product. With his usual clear-sightedness, W.R.Lethaby made short work of the task. The occasion was a public presentation at the Central School in honour of his sixty-fifth birthday (1922). 'What I seem to have found out about life' he gave as follows:

1   Life is best thought of as service.

2   Service is first of all, and of greatest necessity, common productive work.

3   The best way to think of labour is as art . . . By welcoming it, and thinking of it as art, the slavery of labour may be turned into joy.

4   Art is best thought of as fine and sound ordinary work. So understood, it is the widest, best and most necessary form of culture.

5   Culture should be thought of as not only book-learning, but as a tempered human spirit. A shepherd, ship-skipper or carpenter enjoys a different culture from the book-scholar, but it is nonetheless a true culture.

When a man of Lethaby's substance adds up the lessons of his life, it behoves lesser men to be quiet a minute. Could it be that here is a responsibility that the workshops assume, and that the rest of us might indeed take more seriously. For such a redefinition of culture, the artisan approach is in principle a model contribution, in an area of conflict providing one of the deepest problems of reconciliation that confronts our epoch.

During the 1970s the Construction School in the West of England College of Art, Bristol, began to address its energies to the future of the artisan designer. Several workshops have resulted, directly or indirectly. Influential was the successful example of the Goldenhill workshop in Bristol, begun in 1969 by three ex-students from the School. This now well-established workshop does mainly joinery and special-purpose furniture. The development of an educational approach at degree level (only educationists understand what this means) is by no means easy, because, as has been indicated, nothing is further from the hard truths of a workshop situation than the free-and-easy ambience of a college studio. The immediate difficulties are: those of extended reference (what this book is about) to counter the *particularity* of the various production disciplines (whether of hand or machine); the gaining of technical understanding at the right level of complexity; finding the right teachers, particularly those with a broad enough design commitment; and stepping round the worst craft whimsicalities that seem to beset this scale of work. The development of a suitable framework for studies is a long-term project, subject to a good deal of trial and error. It is unfortunate that in the case of Bristol, its development was cut short by the CNAA closing the Construction School, thus losing, in this context at least, the contributions from experienced tutors in the field. (The history of this School is outlined in part 26.)

If unfortunate, then perhaps – other things being equal – to be expected. In Britain, degree-receiving (and awarding) is essentially the dream and the prerogative of the would-be grammar school boy or girl, not to mention their sometimes pushy parents. For this very reason degrees are taken so seriously just a little further down the social gradient. Potting and textiles have always had a strong middle-class following, but otherwise the chap who is 'good with his hands' may be felt to belong down the corridor with the trade apprentices, who are always something of an embarrassment to the higher educationists. ('Good with his hands' – the phrase needs turning over in the mind to receive its full measure.) However, a design-conscious artisan will certainly be uncomfortable with the apprentices; so, except in the very few senior 'vocational' courses, which because of their reduced status carry a penalty for those who teach in them, the design artisan may feel as ill-at-ease in the colleges as in society at large.

As to the transition from college to workshop, Michael Rowbottom has advanced the practical suggestion that students might learn a general theory of machines (and much else) by designing and building their own. Once the materials have been paid for, the machines could then be used in their own workshops, thus easing the awkward problem of starting-capital for beginners. The same could be done for benching, racking, and all the jigs and appliances that a workshop needs. Not all educational establishments would support such notions, and the experimental work to beef them up. Training colleges are often the most prone to doing things 'by the book' but alas they are not alone in this failing. A cynic might suggest that such ideas are too obviously useful and forward-looking to interest educationists. Ivan Illich might add – well, what do you expect; and a harassed administrator might retort that the principles of public accountability cannot embrace the support of free enterprise. Be that as it may, there is now a demand from young people for an artisan-type education, as distinct from a technical training, and viewed as a special case of design education of the normal kind. No doubt a widening of response will become apparent. It is encouraging that more is happening in the secondary schools in Britain; perhaps at the present time the most hopeful sector for change and experiment, as design thinking begins to infiltrate craft studies, and subtly to alter their nature and status.

Artisans of all ages have no problem in finding plenty to read, especially in the technical literature. The 'giant home workshop manuals' are often strewn with useful advice. From a designer's point of view, much of it succumbs to the *false* status and the craft-conscious eclecticism that disfigures so many workshop projects; but, with this reservation, the USA has been a good source for practical manuals of all kinds. With historical references of such high calibre as Salaman's *Dictionary of tools* no one will despair of their bedside reading for quite a long time to come. Three books of wider interest might be mentioned: George Sturt's well-known and evocative *The wheelwright's shop,* Ashbee's *Craftsmanship in competitive industry* –

a fascinating historical document invoking Ashbee's very characteristic notion of 'standard' – and the more recent *The nature and art of workmanship* by David Pye.

In summary, it may be said that the workshop way of life is hard work, immensely educative, often enjoyable, and usually underpaid; it stands on its own feet better than some of the alternatives. It is doubtful if what we may call 'design workshops' will ever organize themselves into presenting a coherent viewpoint, or a representative body of work. There is never much time to spare, even for taking photographs. Nor do such self-observations feel quite appropriate. Perhaps this is the one case where a failure to stand up and be counted is almost enviably reticent. However, good workshops – no doubt rare – are worth seeking out, and occasionally they might offer a useful design apprenticeship.

8   Reading for design

Designers who earn a living by design, seldom have the time to read
many design books. Except – as theatre people understand so well –
when resting. Trade catalogues and the journals can be quite enough.
Thus the main burden falls upon critics and design theorists, who
read each other's books most assiduously. College students and their
teachers struggle to keep up, putting in, probably, the most tenacious
mileage overall; whilst the public at large, users of the things that
designers produce, are understandably less concerned to read about
them. It may be true that book-learning never made a designer;
indeed, the best informed and most articulate people may be the
least directly productive. This is no reason for students to cherish
a defensive ignorance of their subject. The notion that designers
should be noble savages living-it-down in tepee settlements, is not
unreservedly appealing, but even this can be consulted in the
literature. As for intending teachers, it would be nice if they could
distinguish between William, Henry, and Charles Morris, and perhaps
know that Black Mountain was a college (and a good one). Actually
there are enough books and magazines to please most people. Given
the will, an open mind and a good pair of eyes, what is the best way
to set about an investigation?

The following notes develop a simple-minded strategy, in the form
of map-making. By this means, an apprehensive student can hold
the whole situation in the palm of his hand, and decide how best to
serve his own priorities. The authors and titles mentioned are
specimen attractions to be met with in any given area. The selection
has no claim to infallibility, in the way of offering the 'best' or the
latest books – which tend to be around in the studios anyway – but
has two purposes: to body forth some of the ideas that appear in this
book, and to draw attention to some of the older sources that might
otherwise be overlooked. Like buildings, books should be left to
settle for a few years before being taken too seriously. It is fair to
claim that all recommendations are worth, at least, a close look; but
the argument does not depend upon their acceptance. If some books
are out of print, this is not unusual. Design students soon learn that
almost everything worth having is nearly unobtainable ('sorry Sir,

there was never any demand . . .'). Books are no exception, but *all* books are available to those who want them enough. Librarians are usually very helpful. (Part 20 gives bibliographical details of the titles mentioned here.)

Students with broad design interests should consult the mail order list issued by the Design Centre Bookshop, or if in London they will find this shop conveniently central. (See part 21 for the addresses and telephone numbers of this and other shops mentioned here.) The old-established art booksellers Zwemmer's has a wide design coverage, and issues catalogues. Students of architectural design will find their way to the Triangle Bookshop beneath the Architectural Association, to the nearby Building Centre Bookshop for technical books, or to the RIBA bookshop (with the advantage of the excellent reference library upstairs). In the way of general bookshops, a number of nation-wide chains of shops, notably Waterstone's, have come to join the traditional independent booksellers. Now part of such a chain, Dillon's Bookstore (with a specialist art branch in Covent Garden) is the nearest London equivalent to Blackwell's of Oxford or Heffer's of Cambridge. For authors as various as Wilhelm Reich, George Steiner, and Peter Kropotkin, and books and pamphlets on 'alternatives' in all shapes and sizes, Freedom Press Bookshop will be found by the Whitechapel Art Gallery, and the wide-ranging Compendium Bookshop in Camden Town. Regional bookshops, and those abroad, cannot be listed here. Books in print can always be ordered through a local bookshop (who may need your custom) and of course there are very many excellent second-hand bookshops throughout the country, of which the same may be said. College librarians should be a good source of friendly advice. In case of difficulty, students are reminded of the specialist libraries, referred to in part 19 ('Using libraries'), and the other institutions mentioned in the reference parts of this book.

So much for sources of supply. What sort of books is it useful for a designer to read? The answer has to be, that with respect to designing, any sort that helps the job along: that musters energy in the first place, and that then helps to direct such energy where and how it is needed. Obviously this is a personal matter. It may be equally obvious that a designer will ask of his reading, a deepened understanding of his own craft, and of the cultural matrix of its forming – complex and almost unending as such studies can prove to be. However, such reading will certainly include books with no ostensible bearing upon

the contingencies of design practice. Such books may (for instance)
nourish or encourage creative attitudes (or states of mind, of feeling,
sensation, intuition) that are helpfully propulsive to creative activity
in a general way, and to a design commitment in particular; just as
the experience of music or poetry or mathematics might do – not
merely with potency, but with resonance. This might be true for the
sense in which Ernest Newman described an effect of Beethoven's
music: 'It is the peculiarity of Beethoven's imagination that again
and again he lifts us to a height from which we revaluate not only all
music but all life, all emotion, and all thought.' Such considerations
throw into relief the philistinism of training courses (and less) that
masquerade as an education, and whose idea of a 'booklist' may be
somewhat cut-and-dried. However, in their support it must be said
that Ashbee had his apprentices doing physical jerks (pictured in
Craftsmanship in competitive industry) and perhaps he found that
this was just as good. The following notes take a fairly broad view
of what it might be useful to read, and their logic must be left to
unfold toward the recommendations that end the part.

The map proposed has twenty-seven areas, or categories, serving to
distinguish (functionally) one kind of book from another. Books that
elude any such classification can then be considered separately (or
quietly escape attention). The special advantage of what seems, at
first sight, a rather unattractive exercise in category-formation, is that
component books can be added to, or subtracted from, any one
category, if the reader wishes eventually to keep abreast of his
subject. Thus the inevitable 'dating' of any one contribution in no
way impairs the strength of the group – considered as evidence –
which in turn, enables a first-class book that may be getting a bit
long in the tooth, still to have the sympathetic attention that it
deserves. It is easy to become a little patronizing as such a book
slips back into history. A good example is Herbert Read's *Art and
industry*. This book is useful in a minor way to compare Bayer's
typography and layout (first edition) with the tweedy British
aftermath, as has been noted in part 2. However, it is also still the
best available account of its subject, and generally speaking, it is also
the best illustrated. However, the book was first published in 1934.
If Read were alive today he would no doubt tighten up the argument
in a few places (in others, relax it), and he would certainly wish to
reselect a few illustrations of product design. A final – and related –
advantage of the territorial decentralization proposed, is that nobody's
toes are trodden on. If the functional distinctions are reasonably

argued, and useful, then students can confidently find their way around to the point where such aids are superfluous.

One question will certainly be asked (it always is). Why so many architects in a book about design? If the profession is only half as bad as its critics insist – as pompous, as falsely omniscient, and as socially divisive – should we have any truck with them at all? Malcolm MacEwan remarks how well the RIBA building in London expresses all this; and how lastingly. Time has added a dimension of irony to this building's monumental self-regard. Does this mean that in the end, all architecture, like poetry, *must* tell the truth, however hard it tries, so we all get the buildings we deserve?

These are difficult matters. The architects mentioned here are hardly a fair sample of the profession at large (and it is *very* large), and those who write about architecture are hardly more so. However, as A.N.Whitehead said, 'you may not divide the seamless coat of learning', and this is true for the whole development of modern design. It so happens, for reasons that are by no means inscrutable (but which it would be tedious to examine here) that architects have articulated the critical attitudes that belong to that development. Furthermore, in their one-off buildings there was far more scope for experimentation, and for the manipulation of space and form, than has been available in product design. As is well known, however, most of the significant furniture design of this century has come from designers with an architectural training (this is the way the word 'architect' is used throughout this book) so a distinction is unreal at that level.

It is obvious enough that building design provides an accommodating umbrella to shelter many of the smaller-scaled and more intimate design opportunities. From a co-operative standpoint, the question is one of attitude rather than scale (that 'consciousness in common' of which Michael Kullman speaks), unless it is felt that at all costs designers should be prevented from combining (it goes to their heads). For the present, design students risk an improverishment of their work if they attempt to study one design category in conceptual isolation from its neighbours (and parents). At a practical level, the way architects understand manufacture, and close-range detailing, is often stereotyped or painfully vague and abstract; such that architectural students could well profit from a closer experience of the workshops found in design schools. Similarly, whilst the more pedestrian design schools are crassly ignorant of modern architecture, it is also not uncommon to meet architectural tutors who have not even heard of the Bill-Tschichold controversy; far from being aware of its special interest. However, such reservations do nothing to weaken the obvious case for articulating a common critical tradition wherever voices can be found to speak up from first-hand experience; adding such testimony to the work of the professional critics. The life and work of Walter Gropius; what designer's work has not, indirectly, been touched by this man? The remarkable thing is not the commonly accessible nature of one designer's work to another, but that false barriers have persisted so long – and particularly in design education.

1   It remains to add that the worst design books are not usually the work of architects. The 'coffee-table' book is little more than an extended colour-supplement. Such books – large, thick, squarish, and trendy, printed a soupy offset, plastic-covered, and lavish with art photography – such books are a menace to design students. Their imagery reveals nothing that it would be useful to know (like the context, or the cost, or the performance, of the objects they illustrate). Students who rely heavily on this sort of book and the magazines as the full extent of their reading, will hardly be reading these lines, so this category (outside the ranks of the chosen) can be left to speak for itself, which it will do very plausibly. Such books leave the impression of nothingness; that nothing has occurred; but at first acquaintance, a lot seems to be going on.

2   Coffee-table books have their up-market relatives. Like a railway bookstall novelette compared with something by Ivy Compton Burnett, the one is easy and the other definitely sophisticated, but each may be a form of chat. Both categories profit from a willingness to sip, rather than drink deeply of their subject-matter; and their contents are often as out of key as these mixed metaphors. A good example is *Meaning in architecture* (Jencks and Baird), in which some of the brightest English-language theoreticians (Baird / Broadbent / Frampton / Jencks / Silver) *all* have a go, and in which the opening theme is 'semiology'. *High-tech* (by Kron and Slesin) is a rather different specimen. Apart from a certain exclusiveness in their tone of voice, such books may enliven the company of their stuffier brethren, and are even a passable substitute for conversation (in remote places).

3   The synoptic books. These informative and inspirational volumes
bring together what is otherwise apart, by placing designed artefacts
against parallel work in other fields (the arts, sciences, philosophy,
etc.) and in the full context of their origin (for example, a context
of ideas, technology, social circumstance). The extent to which
historical determination is a viable tool of analysis, has come under
question. Books of this type sound large and difficult for beginners,
and invariably are; but they are indispensable in providing frames of
reference, from (or within) which, students can find a personal
orientation and a sense of their own inheritance. The most influential
writers of this kind (such as Lewis Mumford and Sigfried Giedion)
are also, as it happens, the widest in scope. Giedion's *Space, time and
architecture* and its sequel *Mechanization takes command* (to which
can be added the later *Eternal present* trilogy) are excellent basic
books with which to start a design library. As is proper to their
intention, they are marvellously well illustrated (Mumford relies
more on his text). Giedion's work wears well (or, as a boatman might
say, sits well in the water) although in recent years attended by the
corrective nibblings of lesser fry; but he remains, by and large,
magisterially invulnerable within the scope of his large and splendid
volumes. Students might note that *Mechanization* has the only decent
account (to 1989) of modern furniture, now that Gordon Logie's
short book, *Furniture from machines,* is becoming hard to find.

4   The histories. These are shorter, less speculative, more specialized,
than the synoptic works mentioned above. A justly well-known
example is Pevsner's *Pioneers of modern design.* More a collation than
a historical study is Herbert Spencer's *Pioneers of modern typography.*
Certain biographical studies become historical documents in virtue of
subject and authorship, and these (because of their informality) are
worth watching for. William Curtis's *Modern architecture* joins Ken
Frampton's similarly titled book, as recent work of substance.

5   Books on a theme (historical or otherwise): narrowing the focus.
A typical historical document is the facsimile reprint of *Circle* (edited
by Gabo, Martin, and Nicholson), a book that might have been
written last week (but for its optimism) but was in fact published in
1937, to celebrate the return of a prodigal son (the modern movement)
to these shores. Nominally concerned to emphasize '. . . one common
idea and one common spirit: the constructive trend in the art of our
day', these writings seem (implicitly) full of promises: the extent
to which they were kept, is better estimated from a later survey,

The rationalists, edited by Dennis Sharp (1978). Very different is Konrad Wachsmann's *Turning point of building*, a classic statement of the 'only connect' theme at the level of sophisticated hardware (the legendary Alexander Graham Bell steals the show, peering forth from 'a simple wooden shelter in the form of a perfect tetrahedron'). Bruce Martin, our man in junction and modular matters, has *Joints in buildings* and his interesting *Standards and building* as a British contribution.

6   Books about individual designers – Wachsmann (see 5) is almost that – and collections of their work. Examples: the *Life* of El Lissitsky by his wife (*formidable,* as the French might say), *Wells Coates,* by S.Cantacuzino (a book generous of reference), and the many-volumed *Œuvre complète* of Le Corbusier (the best vehicle for the study of this great master short of getting on the road). There is much to be said for in-depth studies of individual designers if the evidence is plentifully available, though, if at all feasible, their work must always be seen and touched as well as read about in books. The danger of 'contamination' by 'influences' can be left to the fantasy life of our smaller-minded pundits. It is useful to correlate observation and reading with critical studies, and 7.

7   Books by designers. At best, there is a spirit and an authenticity in such writing, of an order very different from the efforts of the journalists who follow along after, doing their best to cut the Masters down to size. This is certainly true of Gropius (e.g. *The new architecture and the Bauhaus*), Le Corbusier (e.g. *Towards a new architecture*) and Moholy-Nagy (e.g. *The new vision*). The same is possibly true of Marx and decidedly true of Freud. Original *papers* are now available in collections by historians. An excellent example is Tim and Charlotte Benton's *Form and function.* Here students will find Marcel Breuer's short and very clear exposition of modern movement attitudes, 'Where do we stand?'; a voice from within.

8   Presentation and memorial volumes. Rather a special form of book compiled in honour of some retiring luminary, or perhaps after his or her death. Usually essays contributed by colleagues on any theme of mutual concern, or personal reminiscences; or both. Often such occasions draw forth informal contributions of much historical interest, of a kind that might otherwise be passed over. An outstanding example is *Planning and architecture,* essays presented to Arthur Korn (a well-loved teacher and designer) and edited by

Dennis Sharp (1976). A book as fresh as the day it was compiled and with some exceptionally good photographs. Contemplating the pleasure of these pages, it would be a niggardly spirit that did not feel the sap rising and another spring not too far away . . . A more straightforward example would be Robin Skelton's memorial volume to the critic and poet Herbert Read, forming (incidentally) a good general introduction to his work. (George Woodcock's *Herbert Read: the stream and the source* is worth offering as an alternative.)

9   Exhibits. This category, of books that have 'object-quality' is depressingly small. Any rigour of design consciousness is normally excluded from publishing. Some historical rarities appear in facsimile (e.g. the Bauhaus books). Examples still in circulation as originals will include *Der Stuhl* by Heinz and Bodo Rasch, Tschichold's *Typographische Gestaltung* (which it is instructive to compare with the later translated edition, *Asymmetric typography*), and the oddly charismatic *Flat book* by J.L.Martin. Among late entrants, there is work from Froshaug, from Sandberg (for the Stedelijk Museum, Amsterdam), and from Otl Aicher.

10   Evidential books (see 5). Books that present enough comparative evidence, on any matter of design concern, for the readers to make up their own minds (such evidence would normally include drawings and photographs). A thin entry might be expected here, and the type-form is scarce. Roth's *The new architecture* was a good one. Although confined to photographic evidence, the post-war *Domus* books just qualify. (On desks, fireplaces, kitchens, and so on.)

11   Essays and papers, short monographs. Usually topical, but not always ephemeral; for example Colin Rowe's 'The mathematics of the ideal villa' (*Architectural Review*, March 1947), now collected in book form. *Buildings and society* (edited by King) is a set of essays on 'the social development of the built environment'. Another example: Bruno Zevi's brief *The modern language of architecture*, which is a cheerful antidote to Jencks (see 27).

12   Course books. In the past few years the Open University has been building up course-and-source material of notable quality. Design students (in colleges) may not realize the useful availability of Course Units and Readers, with related audio- and video-tapes, film, and radio and TV programmes. (Lists of these are available from Open University Educational Enterprises, see part 21.) Work by Nigel

Cross and Tim and Charlotte Benton, for instance, is reliably first-class. The nature of OU course-work has varied over the years, in response to public demand and to internal policy-making. 'History of architecture and design 1890-1939' (A305) was a pioneering effort that generated valuable readers and course material. *Man-made futures*, edited by Cross, Elliott and Roy, is a useful anthology, produced for a now discontinued design methods course.

13   Related books. An elusive but potent category. Good examples have been Ruth Benedict's *Patterns of culture* and Jane Abercrombie's *Anatomy of judgement*—the first giving distance, the second, proportion, to any way of studying design. Design being a transactional art, it follows that the looser areas of behavioural psychology have a practical interest, despite the inability of most designers to 'place' such work confidently enough to form conclusions about it. At the sunnier end of the spectrum are books like the catchy *Games people play*, or moving along it a little, authors like Erving Goffman. Designers have always been interested in the Gestalt people – for obvious reasons – and holistic writers like L.L.Whyte have had a similar appeal. Where does a list of related books actually stop? Maybe such books can classify as repositories or tools; the former being heavy and soporific and seeming to sop up energy, like a hot bath; the latter helping to make good work seem worthwhile. A designer's books will (thus) be estimated by their effects.

14   Distant relations. Books that make connections, or present images of critical attitude, in ways likely to be useful. An open and personal category. Some might include *Zen and the art of motor cycle maintenance* (Robert Pirsig); others much prefer their maintenance manuals straight, with pictures. For the literary-minded, Terry Eagleton's *Literary theory* will startle and challenge readers who thought they knew a bit about it. By contrast, the poet Ezra Pound's *ABC of reading* is really a book for writers (and far from dull): arguably an excellent design primer. Title collectors will relish Donald Davie's *Articulate energy* – there is more within. George Steiner's *Language and silence* will not be read by design trendies; but should be.

15   Design philosophy and predicament books. Examples: Toffler's *Future shock*, Papenek's *Design for the real world,* and almost any writing by Buckminster Fuller. For aphorisms in sharp focus:

Frederic Samson's book of sayings, *Dotes and antidotes* (published by the Royal College of Art).

16 Design theory-and-practice. *What is a designer* is one such; John Christopher Jones' *Design methods* an enduring and more specialized example. The latter-day Chris Alexander is tirelessly and biblically prescriptive in his series beginning with *The timeless way of building;* with Bachelard, a literary man's view of formal values.

17 Technical books generally. Necessarily a mainstay of any designer's consultative reading. Too many and various to list or describe. Such books will be specified by course tutors. (See also part 18, 'Finding out for yourself'.)

18 Catalogues. Possibly a sub-set of 17 but rather different in kind and function, and often more ephemeral. There is no doubt that catalogues claim an unchallenged primacy in the field; some designers read little else. Catalogues may be related to the big trade exhibitions, or to wholesale distributive sources, or may be directly and concisely informative about a particular range of products. A few, like the Wilkes-Berger, are splendidly suggestive. It has been well said that any man in a tight spot will draw comfort from these pages; yet, it is not easy to describe the special satisfaction of well-presented ironmongery. Then there are the great catalogues; always, alas, of the past, and still to be chanced upon, occasionally, in country houses, or upon the bookshelves of retired shop-foremen. To scan these is to enter a lost world of mechanical authority; the steel engravings alone are of a privileged artistry. The Guillet machine catalogue might be spoken of in this context; but rarely seen. Catalogue *systems* are another thing altogether and appear under 20.

19 Definitive references. The major works, like Needham's *Science and civilisation in China,* or the big encyclopaedias, or the OED.

20 Common references. These include the technical reference systems, using CI/SfB classification (commercially serviced or run by the design office) and the standard reference books such as *Specification.* Less obvious are publications like *Which?* as a consumer check on the retail market, and the small handy books (including diaries) that rank as pocket references. These might include (for instance) Matila Ghyka's useful *Geometrical composition and design,* or Charles Hayward's *Woodworker's pocket book.* For English-language

purposes, designers may well use Chambers, Fowler, and Partridge (*Chambers twentieth century dictionary,* Fowler's *Dictionary of modern English usage,* and Partridge's *Usage and abusage,* respectively). See part 18, 'Finding out for yourself'.

21 Books on education. Not all students are too keen to read about education while still in its grip; yet it is a proper function of any educative process to question itself and informed questioning is the best kind. Those who wish to take a critical backward look at their schooling might enjoy Reimer's *School is dead,* Goodman's *Compulsory miseducation,* and the irrepressible Ivan Illich's *Deschooling society.* A centre-view is provided by R.S.Peters (whose 'philosophy of education' is attacked from a Marxist angle by Pateman in *Counter course*) and by A.N.Whitehead's trenchant *Aims of education.* On the left the legendary fieldworker A.S.Neill has *Summerhill*; on the right are Cox and Rhodes Boyson in the *Black Papers* (not always quite as black as they sound), and perhaps Bantock.

22 Books on design education. Many omissions. For instance, it has been hard work to get the facts about the Hochschule für Gestaltung at Ulm. The international disturbances of 1968 have been well documented; for England see the Penguin *The Hornsey affair* and Tom Nairn in *The beginning of the end* (evidence from the other side appears in the *Report from the Select Committee*). Educational politics grind on remorselessly, with a few specialists attempting documentation and the occasional challenge. Gordon Lawrence and David Page attempted this in their essay in *A degree of choice* (edited by Finch and Rustin). The Bauhaus is now well covered: given a good library it is worth going to Wingler's *The Bauhaus*; failing which the Bayer/Gropius *Bauhaus* is probably the best short survey. There are now an increasing number of books on design education at secondary school level (a matter of interest to many students taking degree courses, apart from its intrinsic merit). An example, *Design education in schools,* is by Bernard Aylward, who pioneered the Leicester experiment, which combined course development with new school buildings adapted to it.

23 Magazines. British design (and design/architectural) magazines seem to proliferate endlessly; they are well enough known in college libraries and specialist shops; overseas equivalents will be found there too. It is less well observed that all trades have their

magazines, and these are fruitful sources when required (even for their advertisements). Some magazines generate books out of special issues or series of contributions (the *Architectural Review,* for example). Among recent enterprises, the English *Form* was a welcome if brief visitant; worth looking for. Good magazines of a quite unconnected kind will sometimes have a run of a year or two when their quality is exceptionally high (and therefore of interest beyond the charmed circle of their ordinary readership) – this happened to the magazine *Anarchy* under the unobtrusive editorship of Colin Ward, and more recently (1979-80) the American *Wooden Boat.* It would be hard to connect these with each other, or either with a designer's normal interests (so watch out).

24 Bran tubs and oddities. These are books with something for everybody, and the *Whole Earth Catalog* is surely the most kaleidoscopic. This one is as hard to classify as it intends to be. An example of an 'oddity' (only by the present terms of reference) is *Richard's bicycle book* by Ballantine: handy for those with bicycles, obviously, but also exemplary for its illustrations (designers please note) and for the agreeably unpretentious – yet informative – nature of its discourse (other technical writers might note).

25 Alternative attitudes. Well-known examples are the diagnostic *Blueprint for survival* (by the editors of *The Ecologist*), and E.F.Schumacher's heart-warming *Small is beautiful*—with its sequels *A guide for the perplexed* and *Good work*. Fading, but not without interest are Roszak's *The making of a counter-culture* and Guinness' reply to it *The dust of death*. Of constructive account beyond its strict subject-matter (housing) is John Turner's contribution as an activist and field-worker, brought together in his book *Housing by people,* putting the case, baldly speaking, for self-build and for the social principle of user-control. On a similar wavelength are books by the architect Colin Ward (e.g. *Vandalism* and *The child in the city*), who edited the *Bulletin of Environmental Education* and directed the Schools Council's 'Art and the Built Environment' project. Mike Cooley, the shop-steward from Lucas Aerospace, provides a radical reappraisal of computer-aided design in *Architect or bee?* Women's movement contributions include *Making space* by the group Matrix, with a useful bibliography.

26 Alternative hardware. A rapidly growing collection of books on all aspects of 'loose fit, low energy, long life' (Alex Gordon's

formulation, see *RIBA Journal,* January 1974) and how to do it in the design and building of devices, mechanisms, and environments of all kinds (what the motorcycle sidecar enthusiasts used to call 'the outfit'). As would be expected, advice ranges from the homely to the arcane, and from DIY technology to advanced engineering. Students wishing to study wind power or solar energy systems, for instance, will find a considerable literature awaiting them. Walter Segal's refreshingly down-to-earth designs for low-cost housing were reviewed in the *Architects' Journal* (and see the special memorial issue, 4.5.1988).

27   Anti-modern movement. Currently as fashionable as plug-in technology was in the 1960s and taking one of two characteristic positions : the first, that there is an identifiable body of work now sufficiently considerable to be called 'post-modern' and that the 'modern movement' which it replaces may now be deemed a spent force; or the second, that the modern movement had no real warrant for the largeness of its claims, has got itself into a formal and philosophical cul-de-sac, and that this is consistent with the faulty premises of its origin. The first position is illustrated in various books by Charles Jencks (such as *The language of post-modern architecture*) and the second perhaps best by *Morality and architecture* by Watkin. This book and others similar are discussed in a critical symposium 'The edifice crumbles' (*Architectural Review,* February 1978) to which interested readers may wish to refer. The controversy is less simplistic than the above summary might suggest, but somewhat scholastic in tone and stylistic in its central interest. The anti-*design* writers are far more radical, and having rejected the design profession as it currently professes, are more concerned to adduce alternative attitudes (see 25). Their rejection follows from the view that decision-making generally in the community needs to be deprofessionalized, involving far more user-participation than is now the case, and that designers (when their concerns are not wholly trivial) may in the end do more harm than good. Turner and Ward seem on the whole to take this view, though it is entailed by their position rather than central to it. Malcolm MacEwan's *Crisis in architecture,* easy to group with the writings of those who decry the modern movement, is in fact a perennial reminder, and in this case a constructive one, that all is not well at the RIBA.

This completes the suggested *range* of possibly useful design reading, without closing the door on late comers, new arrivals, or other likely entrants. It must be said once more, however, that for a pot, a building, a rug, or even a well-baked loaf of bread, no amount of reading will take the place of touch, sight, taste, sound, and smell; by which token it should be added that the only reliable form of words is a poem, and possibly a novel – all other forms being insufficiently transparent. We read criticism to increase our enjoyment of these things, to understand them, and perhaps to make them better. They are, after all, leavings.

As to the modern movement and how is it doing, the answer this book gives is – not very nicely thank you, but as well as can be expected, taking everything into account . . . Designers do, for practical reasons, inhabit a house of theory of their own; it concentrates their minds or helps to. These notes have been concerned to open doors and windows, and to welcome guests. The structure is surely a makeshift one, draughty, and unsure in its foundations. This is true of all such structures since Hiroshima and the concentration camps : the extremes of contemporary reference.

It used to be said that to keep sane, modern man must either come to know God, or be in permanent urgent conference with Marx, Freud, and Einstein, or come to an arrangement with the Devil and go into business. Few designers have lost themselves wholeheartedly in any one of these alternatives – if they had any serious interest in the modern movement. Perhaps they should have done, and perhaps they failed to learn deeply enough the lessons of dematerialization, the *reductio ad absurdum* of their reductionism; being thus neither scientific enough nor religious enough. After all, Beethoven, who on any sensible estimate of 'the present' must still be accounted our greatest contemporary, made matters plain in his late piano music (and perhaps in the quartets for those who have caught up with them). W.J.Turner wrote an interesting little book to adduce the view that 'all art is the imagination of love, and music is the imagination of love *in sound*'. The parallel suggestion is that 'well-doing' in Lethaby's sense is the play of love upon inert material; the finding of its forms – to which the clarifying tasks of our epoch, the finding of each other, and the conversion of mass into energy and relationship, are contributory. In a situation of physical entanglement (the world of the transactional) there are two images of correspondence to such tasks: that of definition, the 'area of concern', and that of conduction, 'energy striking through' (the through-and-through requirement of modern design, which gives a characteristically figure to ground situation in which the perimeter or frame is unacceptable). Christians who happen to read these lines, or between them, will immediately see what is implied: it is indeed possible to argue that the modern movement *an sich* was and is a making-straight the path of the Lord; is about and around some implications of that task; and that where it fails is always the mark of secular presumption. It is amusing to think how such a speculation might be received in some quarters. Related studies, the concern of this part, might wish to endorse or to dismiss such a claim – but in either case by their fruits let them be judged; for such studies will only be helpful if response is quickened, and life and love allowed to become more abundant. It is well known that where apathy dwells, comes the evil day . . .

Categories, however, are clusters, and this part will end with a linear connection. It is a cheerful one, and demonstrates a further dimension of critical resource in the modern movement. If it can be argued – as indeed it can – that the history of consciousness is a chain of hands, then our twentieth-century situation demands that the hands we seek shall be neither flabby, evasive, nor treacherous. 'Good with their hands.' Where shall we look?

Among the founding fathers of the modern movement, which description must include William Morris pre-eminently, let us begin with his friend and contemporary the great Peter Kropotkin, author of *Mutual aid: a factor of evolution* and the prescient *Fields, factories and workshops* (recently re-edited by Colin Ward). Kropotkin (1842-1921) was a source of inspiration both to Patrick Geddes (1854-1932) and later to Lewis Mumford. In the first decades of this century there are three writers – all active and substantial men – who might be said to stand for the concerns discussed here: Geddes, Ashbee, and Lethaby, who were approximate contemporaries. Geddes, splendid man that he was, is usefully appraised through Paddy Kitchen's biographical study *A most unsettling person* (there is a lengthier book by Boardman). He was a synthesizer of ideas and a social activist, much concerned with user-participation; in which regard it must be clear that men like John Turner and Pat Crooke (or Illich) stand within the tradition of concern of the modern movement, and not, as some might suggest, outside it. Ashbee said of Geddes 'his work has ever a touch of prophecy' and 'when all's said and done . . . his prophecy is likely to sound the farthest'. It is interesting and rather unexpected that he found no hesitation in commending Mackintosh for his Glasgow School of Art, saying 'the real artist is he who, like Mackintosh in the Art College of Glasgow (one of the most important buildings in Europe) gets his effects within the sternest acceptances of modern conditions. For here never was concrete more concrete, steel more steely . . .'

Ashbee was an interesting man whose concern with the workshop concept, and what might be done with it, is perhaps not as well known as it should be. (See his *Craftsmanship in competitive industry,* an account of the problems faced by his Guild and School of Handicraft [founded 1888] in moving from London's East End to Chipping Campden.) He was, as Pevsner implies, no intellectual Luddite and stated in *Should we stop teaching art?* (1911) that 'Modern civilization rests on machinery, and no system for the endowment, or the encouragement, of the teaching of art can be sound that does not recognize this' – a shift both from Morris and from his own earlier position when he spoke with rather more reserve: '. . . we do not reject the machine, we welcome it. But we desire to see it mastered . . .'

Lethaby is in some ways a larger and more recognizable figure, partly because of his contribution to design education as Principal of the Central School of Arts & Crafts in London, and as a Professor at the Royal College of Art; but better, perhaps, for the vigour and clarity of his writing, of which his book of essays *Form in Civilization,* reprinted in 1957 with an introduction by Lewis Mumford, is an excellent example. Lethaby seems to gain in stature as he recedes into history. He has been called a moralist (is this a human failing? he was certainly no prude or killjoy) but he is better seen as a realist, impatient with aestheticism and what Mumford calls 'the shams and sophistications that Lethaby loathed'; always, however, with a moral dimension at the core of his thinking. Here are some characteristic aphorisms:

Art is a natural human aptitude which has been explained almost out of existence.

The task of civilization is to add to what may be loved.

He owns most who loves most.

To live on the labour of others is a kind of cannibalism.

There are too many professional men, tradesmen and noblemen: too few men.

Lewis Mumford finally met Geddes in 1923, finding him (as was common) extremely difficult, though his presence 'shook my life to the core'. Mumford himself became one of the wisest synoptic writers of our time, influencing (together with Giedion) several generations of design and architectural students through his books *Technics and civilization, The culture of cities* and *The condition of man*; he is still, of course, active (1980) and has been a good friend to students and designers who search continually below 'the surface of the face of things' for the meaning of our time, and how they should best respond to its demands.

Mumford writes from America; in Britain, probably the last critic of the breadth and sensibility to carry forward this tradition was the poet and essayist and polymath Herbert Read, who has already been mentioned for his book *Art and industry*. Read, who was also a follower of Kropotkin – more explicitly so than Mumford – wrote widely on the arts, philosophy, sociology and literature. His substantial contribution *Education through art,* which did so much to influence British primary school teaching, was warmly admired by Walter Gropius. Read's personal impact upon British cultural life is no longer well remembered – he is too near – but in his modest and seemingly tireless way he continuously urged and promoted an awareness of modern art and design (and literature) that was slow in coming to these insular shores from the continent and America.

Here, then, are joined hands from Kropotkin through Geddes, Ashbee, and Lethaby, to Mumford and Read, and on to E.F.Schumacher and others like Colin Ward and Paul Goodman; and here, observably, is a tradition of warmth and human concern in a critical culture, linked at every point to the growth of the modern movement, and emphasizing its roots in the *necessary* thinking of the twentieth century in respect to its complex inheritance. It would be tedious to develop this indication – it is no more – into possible cross-connections, which it may interest students to do; nor should a world of social imperatives be found coterminous with that of nuts and bolts and object-making, the 'traditions' for which are outside the scope of this book to examine. However, Instant Trend and Moneyman (and other television stereotypes) must be contained in their redefining of our culture. Those who are exasperated by the confusion of stylistic trivia with the spirit and attitudes of the modern movement, which are the source of its relevance, might draw encouragement from some of the sources mentioned. For my part,

these are hands I would take in complete confidence. Like the poor
(it seems), what is modern is always with us; and open to
transformation. Here is Herbert Read's poem 'A song for the
Spanish anarchists':

The golden lemon is not made
   but grows on a green tree:
A strong man and his crystal eyes
   is a man born free.

The oxen pass under the yoke
   and the blind are led at will:
But a man born free has a path of his own
   and a house on the hill.

And men are men who till the land
   and women are women who weave;
Fifty men own the lemon grove
   and no man is a slave.

9 Summary: students as designers

Students who want to be useful in the world, who see chaos and want order, who lack self-confidence, and who are swamped by the apparently required armoury of skills and facts, are sometimes rashly seduced by unitary views of disparate phenomena – it seems hopeful that way, and it seems manageable. Those who court lateral thinking to infinity can easily arrive at the semantic double-think in which concrete equals abstract; an achievement which some design theorists have managed effortlessly. Many a pretty paradox can flourish, and indeed be artfully cultivated, in a society that lacks a spiritual dynamic, and would like the best of east and west without the penalties of either.

Design has a very broad spectrum of opportunities that can be made socially worthwhile. Some are mechanical, some ephemeral, some interpretive, some call upon unique solutions, some move toward those generalized type-forms that Gropius talked about. Designers themselves are equally various in personal make-up. The first task for a student is to know – and accept – himself. The second is to educate *himself* (and here, because the word 'himself' has special weight, it is a relief to break the language convention – a problem throughout this book – and add *herself*); accepting formal education as one set of constraints and opportunities exercised within the temporary benefits of community. Remembering, in this matter of education, a saying by Robert Graves – that it is easier and more common to hate hypocrisy than to love the truth. The third task is to sense out, quite intimately, a growing sense of accomplishment that is accessible simultaneously to eye, hand, reason, and imagination; testing each ground to spring from.

This is not an original specification for survival. The point to reiterate is that so few survive at the level of creative expectancy that formal education generates – and to which subsequent experience adds, too frequently, a tepid aftermath. As things are, the one thing about the future of which a student can be sure, is that its demands on him are strictly unpredictable: the rough shape of possibilities may be discernible, but the exact and tangible nature of a creative challenge is a happening and not a forecast.

How, then, to keep open to the future with enough personal acumen and buoyancy to cope with its opportunities at full stretch? If students feel blocked by society as it is, then they must help find constructive ways forward to a better one. In a personal way, the question *must* be answered by individual students in their own terms, but as far as design goes, it is possible to see two slippery snakes in the snakes and ladders game. The first snake is to suppose that the future is best guaranteed by trying to live in it; and the second is an assumption that must never go unexamined – that the required tools of method and technique are more essential than spirit and attitude. This snake offers a sterility that reduces the most 'correct' procedures to a pretentious emptiness, whether in education or in professional practice.

The danger is reinforced by another consideration. There can be a certain hollowness of accomplishment known to a student in his own heart, but which he is obliged to disown, and to mask with considerations of tomorrow, merely to keep up with the pressures surrounding him. Apart from the success-criteria against which his work may be judged, there is a more subtle and pervasive competitiveness from which it is difficult to be exempt, even by the most sophisticated exercises in detachment. Hence the importance of recognizing that education is a fluid and organic growth of understanding, or it is nothing. Similarly, when real participation is side-stepped, and education is accepted lovelessly as a hand-out, then reality can seem progressively more fraudulent.

Fortunately, the veriest beginner can draw confidence from the same source as a seasoned design specialist, once it is realized that the foundations of judgement in design, and indeed the very structure of decision, are rooted in ordinary life and in human concerns, not in some quack professionalism with a degree as a magic key to the mysteries. From then on, to keep the faith, to keep open to the future, is to know the present as a commitment in depth, and to know the past where its spirit can still reach us. As to know your enemy is more usefully to know, and to seek, your friends. The enemy can usually be trusted to reveal himself (or is only too easy to manufacture).

A designer comes to recognize that his world is only fleetingly conjectural. Just as a student finds it hard to believe that anyone will actually give him a job – and there he is five years later, working and still managing to eat – so it is with all our legendary tomorrows: they offer their problems concretely enough, if we are *there,* and refuse to take no for an answer. It is also true that a designer interprets reality through the modalities of action; in the end, his work stands or falls by the intractably objective qualities of an outcome. Only in such good sense is he a philosopher: in making actual an experience of design, and thus constantly redefining what the word (and the work) stand *for.* Yet because we occupy our differing roles only in virtue of our common predicament, which is our humanity, there is no end to the questions we must continue to ask.

Design is thus simultaneously a realm of values and a matter of engrossingly particular decisions, many of which are highly technical. There is a threshold up to which we can quantify, and this is often enough the task for a professional: less an equation of meaning than one of ordered evidence. Beyond this threshold, design is strictly a cultural option. It always has been. We humble ourselves, we sharpen our wits, and we offer, at the very least, our moments of lucidity. Our concern is always 'the place of value in a world of facts', but there is no role waiting for us, there is merely the chance of making one out of the sheer courage of our perceptions. In the same way, if you want to link hands with the spirit of the modern movement, it won't come to meet you; you must go out and make it your own.

Notespace

10   Explanation

The change of paper indentifies the reference area of this book, and
should make it easy of access. Although the notes may be consulted
separately under their headings, the *key parts* are 11 and 12 ('How is
design work done?' and 'Communication for designers'), which put
the rest into context. A simplified overall picture is the first
requirement. The detail can be filled in by experience, or – when
needed – from the more specialized textbooks. For this reason, and
to keep the whole thing manageable, the notes are kept to basics and
essentials. If the practical advice and check-lists are prefaced by
general discussion, the advice will be found restated in tabular form
at the end of the part. Where more specialized work is touched upon
(for example, that of a surveyor in relation to the survey notes given
in part 15) the advice is reduced to considerations that all designers
will find essentially reasonable. At a more general level, some aspects
of diagnostic technique are seldom given the attention they deserve,
and will seem unfamiliar for that reason; for instance the brief
notes on 'Asking questions' may seem quite difficult to anyone
without training in the field. Practice becomes the answer.

It is hard to strike a balance between professional realism, and the
more personal and informal design situations. 'How is design work
done?' gives too simple a picture for building design, yet is far too
elaborate for textiles or ceramics. However, designers must know
how things happen in the ordinary commercial world; the conventions
employed, and the order of events as normally experienced – whether
in lesser or greater detail. It is against the demands of the transactional
world that technique has to be evaluated. There is also some
correspondence between 'the way things go' and the various possible
logical models of the design process, which (when they work) may
offer some help in allocating effort where it is most needed.

This matter is discussed in part 6, 'Problems with method', and is
further examined here – in a practical way – in parts 12 to 17
inclusive. However, the recommendations are chiefly concerned to
develop technique in harmony with understanding, and stop well

short of any more systematic view of design methodology. The difference may be seen on a rough-and-ready workshop analogy, for which 'method' would represent an overall strategy for all cases, how things are best put together and in what order; 'technique' being what tools to select, how to keep them sharp, and how to use them. Both have their place, but in a workshop sharp tools are the first requirement. Both are most appreciated after mistakes have repeatedly occurred and much time and energy been seen to be wasted (on the workshop analogy, from muddle and from blunt tools). Thus students may find these notes a bore, or a mystification, in the first year, but thereafter and progressively, they should prove helpful. As is so well known in sports training – or in any serious pursuit of high standards – technique is hard work, and this is inescapable; there are no short cuts. What of method?

During the past twenty-five years many design schools have experimented with different aspects of problem-solving disciplines and with systematic design method. Before 1968 – a useful watershed – substantial work had come from the Hochschule für Gestaltung at Ulm. In Britain there had been notable published work from Bruce Archer and John Christopher Jones. David Warren Piper surfaced at Hornsey College of Art during the troubles, with some interesting work. My own contribution at the Royal College of Art had to do with diagnostic technique, and communication methods, briefly summarized here. Chris Alexander wrote the suggestive *Notes on the synthesis of form* and Jane Abercrombie her very useful *The anatomy of judgement* and many papers on learning and teaching derived from her research programme at the Bartlett School of Architecture and elsewhere. In America, the prestigious Buckminster Fuller was still soldiering on, in pursuit of global strategies for survival and an 'anticipatory design capability' to put them into effect.

Since 1968, designers have tended to take a less messianic and more piecemeal view of their social role, with a bit less confidence in the grand strategy. This is perhaps reflected in methodological studies, which apart from advance in the field of CAD (Computer Aided Design) have been concerned with retrenchments and reappraisals. The matter is nicely summarized by Nigel Cross of the Open University in his paper 'The recent history of post-industrial design methods' (1979), which also has a useful set of references. Interested students should also consult J.C.Jones' *Design methods*. For a more detailed account of professional practice and office procedure, see

Dorothy Goslett's *The professional practice of design* and Ron
Green's *Architect's guide to running a job*. J.B.Creswell's classic
Honeywood file is a cautionary tale; very amusing, readable, and
instructive.

11 How is design work done?

Much design work is carried out in a very direct and informal way. The degree of formality becomes a function of scale and the number of interests represented. Of course an artisan designer (of any kind) works very directly and with a minimum of 'communication procedures'. As was explained in part 1, however, this book takes a middle-range view of design opportunities and here matters of procedural technique become critical, for reasons given in the following parts.

The step-by-step account of events that follows, is intended to place design procedure in its normal context for (say) the design of the interior of a small department store, in this case to include some structural work, all fitments, and graphics; or that of a large exhibition, which would be comparable. The sequence is 'normal' for present purposes, but naturally there are variants.

From such an account it is possible to abstract out an analytical model of the design process, and thus to see what is irreducible over a wider range of problems. This would merely confuse the descriptive purpose of these notes, but readers are directed to methodological references (see part 10) for a systematic view in such terms.

The procedures show the designer approaching an unknown situation, making himself familiar with it, taking instructions, making sure they are fully understood, weighing the possibilities, discussing them, arriving at conclusions, offering proposals, modifying them, providing drawings and other instructions to a third party, and supervising the outcome. The result is something new in the world; a product, an environmental change; a new set of possibilities. In dynamic balance to effect this change are the interests of four persons or groups of people: the owner or client or employer (the word client is used because, for the designer, it is the least ambiguous); the user or public or consumer; the designer plus any technical advisers; and the contractor or manufacturer. Each party deploys social resources in a different way and there is no parity of representation between these interests; that of the larger one (the user or public) often being, in effect, mute. There is, however, an ongoing economic relationship between the client and his market or public, which has to some extent its own system of checks and balances: the designer and contractor come into the situation briefly as agents of change, with the designer most potently concerned with its nature. His responsibility is therefore a complex one: is he working for himself, for God ('good work'), the client, the builder or manufacturer, the public or user, or in some embracing respect the interests of society as a whole? He may feel that the resources available (labour, skills, materials, and including the special skills of his client and his own professional training) are a social resource, and that their effective control through money at the disposal of the client, is largely an arbitrary matter; that his role is simply to free and 'optimize' an outcome as best he can; or he may take the view that his direct contractual responsibility is to his client's personal interests which he is called upon to serve, and all else must be so governed. The designer must be aware of the highly contingent nature of all his decisions in these respects; he must also remember that each possibility is a new one.

1   Letter or telephone call from client or equivalent
    Arrange a meeting at his and your convenience.

2   Meet client
    This is an unpredictable occasion, though the client will usually be anxious to convince himself that you are competent, experienced, personable, and able to look after his interests. Prepare accordingly. If you decide to take some work, a vast and shapeless portfolio may prove damaging by making the client feel that you are setting the pace, and in doing so, exploiting his own lack of technical expertise. Equally it is unwise to turn up with batteries of assistants, tape-recorders, and measuring tapes, etc. It is most essential to *listen* properly, and to take notes; listening being that part of conversation that a designer must practise and get word-perfect. Briefs; see 15.

3   Visit site, meet other interested parties
    This is again an informal occasion – not a site survey – but an important one, because first impressions are important. Note them. The 'parties' may be the clients (plural) or the client's business associates. Take camera, but only use if the occasion seems suitable.

4   Exchange of letters (leading to contract or letters of contract)
    Must be considered in formal terms; free of too many references to

'I', and not over-burdened by design jargon or unsuccessful attempts
to sound like a business man (e.g. 'the favour of your letter' and
'may we assure you of our best attention at all times'). The best form
usually is a direct and friendly letter, practical and reserved in tone,
with facts or lists separated out as an accompanying sheet, such that
the client can hand this round without his colleagues having to work
through the more ephemeral or personal bits. At the beginning of a
design job, it is often helpful to a client to have *a very simple* account
of the design sequence that you usually follow, so the client knows
what to expect and roughly when to expect it. Be careful, however,
unless the job is one you have encountered before. A letter is a good
vehicle for this simple information. You should advise a client that
consultants may be necessary (engineer, quantity surveyor, etc.) and
you should make him aware that formal consents may be involved,
and that it may be necessary to consider his personal insurances, his
obligations of tenancy, the rights of adjoining properties, planning
and fire office consents, and (in the case of an exhibition) the
exhibition regulations.

Obviously, correspondence continues to the sweet or bitter end of
the design process.

5   Contract or letter of contract
    Legal documents, highly conventionalized. Should be separately
    studied as an aspect of 'professional practice', the subtleties of which
    are outside our present terms of reference.

6   Letters of record or enquiry to other interested parties
    Such letters will always be short and to the point – such letters are
    almost conventions and most of them go into the filing records of
    local authorities, etc.

7   Site and premises survey, subsequently drawn up
    It will save endless work if this is done properly because the drawings
    must be very accurate descriptions of fact, accompanied by
    photographs, recordings if necessary, and extensive notes. Sometimes
    you will employ a surveyor to produce the measured drawing; you
    will still need your own observations for design purposes. See part 15,
    'Survey before plan'.

8   Questions
    The asking of questions, and finding the right ones to ask, is

fundamental to such dignity and relevance as a designer's role can conceivably posses; this process is no conjuring act or trick of the trade with intent to deceive; on the contrary, the purpose of questioning is to identify and solicit such truths as may be available. See part 16, 'Asking questions'. Here, it should merely be noted that questions may be asked orally, in which case it is sometimes acceptable at this stage to use a tape recorder, or questions may be submitted rather more formally as a questionnaire. Questions will include requests for factual information available to the client (e.g. for a shop, very elaborate questions about stock and sales policy). It is unwise to ask questions casually by letter or at random meetings. The client will usually respond well to carefully prepared questions; they are a measure of your concern for his interests.

9   Research, permissions, consultation
    This will involve a consultation of all other parties or interests involved, particularly local authorities in the form of the district surveyor, fire and health officers, etc, together with all necessary technical information (catalogues, etc.) and anything else necessary to complete your reference data. There may at this stage be any relevant form of public consultation, which can occur through representative bodies or pressure groups, or by the summoning of meetings, or by taking samples of local opinion, or by the more controlled 'opinion poll' method, or by seeking available evidence from past experience. It is doubtful if any of these approaches would be employed for the scale of work under discussion (though they could be), but it is necessary to mention them here. Research and consultation is a considerable and continuing task and is not neatly confined to this stage of the job.

10 Sizing up the job
    At this stage you should have enough information to work out an office programme with a clear allocation of personal responsibilities for the conduct of the job. This exercise will give you an intuitive idea of the possibilities latent in the design work (conditioned by time, money, and the nature of the problem) and just how far you can afford to develop them. This will affect your whole subsequent approach; your personal strategy.

11 Preliminary ideas
    Design begins here in a formal sense, though you may by now have a design concept or a 'mental set' from your first meeting with the

client or your first view of the site (this is a matter of experience). Mistrust mental sets until they have proved their relevance. This is the stage at which ideas are roughed out diagrammatically and working principles examined, tested, and agreed. Such ideas may take a mainly visual form (but will usually be diagrammatic in character) or, if you use a 'report' as a kind of developing discussion with yourself, may be verbal argument accompanied by concept diagrams. It is difficult to generalize about the thinking that may go on at this stage.

12   Report
A report is not always necessary, but it is a most valuable instrument for two-way consultation. See part 17, 'Reports and report writing'. The report will embody your proposals in principle and your reasons for suggested courses of action. The report may include diagrams and a plan layout, and may include catalogues and other references. The report should *not* commit you to what the job may finally look like. You should normally include a discussion of alternatives, for three reasons: to show why your proposal is the best one (as it should be), to make it clear that all possibilities have been adequately canvassed, and to gain 'reasonable assent' to what you propose, which is best done in an open contemplation of the options. On the other hand, it is unwise to present evenly balanced alternatives, unless you are really convinced that it is right to do so (your client may think, 'what the hell am I paying him for?'). The report is *not* a brief, however, one of its purposes is to focus discussion, and another, to bring to light new considerations which your client, or other interested parties, may not have thought of hitherto; finally, its purpose most succinctly, is to make the definition of a brief securely possible, and less liable to be faulted. Always remember therefore that a report is a consultation document.

13   Pause: recipients consider your report, copies of which you may have sent by post or presented personally.

14   Meetings, conversations, etc: to discuss the report and your client's reactions to it.

15   Brief
A 'brief' is a statement of your agreed terms of reference for developing a design to presentation stage. In a small or simple job the brief will emerge much earlier, in correspondence with your

client. Here, the brief emerges from final agreement after the report has been fully discussed, with all modifications taken into account. It is highly desirable to set out your brief in a manner that obviates any kind of misunderstanding, but it is by no means an onerous task (not comparable in length or scope or detail with a report). Its purpose is to reassure the client that all his reservations have not been brushed aside, and to remind him that from now on work must be seriously under way without his interference. This is the last chance for the client to reject your approach to his problem (unless he finds your presentations drawings unacceptable).

16   Development

Here begins 'design' in the wholly conventional sense. You will have a model from your survey; if not, make one now. According to the nature of the job, you will proceed on the drawing board, in the layout pad, and in your workshop (if you have one) examining and developing ideas and testing them against your problem analysis if you have one as such, or against the reference data you have accumulated from your client and other sources (or against your report if you used the report as an analytical tool). You will discuss alternatives with trade representatives, with your quantity surveyor if you have one, with any relevant local authorities, and you will form a good idea of the cost involved. All this will lead to a 'presentation' of your ideas, or rather your intentions, to the client.

17   Presentation

This will normally involve a model, sample sheets, diagrams, notes, sketches or 'perspectives', graphic layouts in whatever form appropriate, and will be personally 'presented' and argued for by yourself to the client or more usually to his board of directors. You may need an assistant, and, if it is a combined graphic-construction job, both of you will be present. A design presentation should go pleasantly enough *if* the work outlined above has been carried out properly. A presentation should be seen in definite terms (if it is inescapable) and any alternative ideas should merely be supporting evidence for your proposals.

18   Modifications

Certain changes may be necessary in the light of your client's response to the design presentation. Normally, such changes can be agreed informally (covered by letters) in your own office, and will not involve the effort of a further presentation.

19 Working drawings

The largest task now begins. Working drawings, or construction drawings, are detailed accurate informative *instructions* on which the work will be carried out. This is too complex to discuss here, except to say that drawings for consent will normally be done first, and the remaining bulk of the drawings must be carefully planned for the intended contractor or sub-contractor or manufacturer. Drawings may be submitted to a quantity surveyor in order to receive a bill of quantities (a measured account of all labour and material items involved in the job) or may join a written specification and schedule in submission to competitive tender. These are technical matters which need not worry you in this short summary. Discussions with contractors may involve many modifications as the work proceeds, but if the contractor is unknown (i.e. to be decided by tender) such modifications will ensue after the contract has been settled.

20 Tender

Designers (in this case, as distinct from architects) will often work in an informal way with a contractor already well known; in which case much of this procedure loses its worrying aspects, but inescapably, however the job is done, a very large number of drawings must reach the contractor with sufficient notes or verbal description (as in a specification) to leave no room for doubt. (A specification describes in words the quality of materials, workmanship, etc, and otherwise complements drawn information.) Students may hardly be concerned with these procedures and this part of the working sequence has been simplified accordingly; for a more detailed picture, consult the references suggested at the end of part 10.

21 Contractor begins work

22 Site supervision

This is an important part of design procedure. With the best will in the world, and a very good set of drawings, things will always go wrong on site and last-minute alterations will be essential. The designer must also make sure that deliveries are properly in hand. It is best to confine site instructions to the site foreman acting as a general co-ordinator. Visits will also be necessary to the joinery shop, to the printers or typesetters or indeed to any supplier or sub-contractor.

23 **Payment**
The designer will normally certify contractor's claims for submission to the client. In the case of shop-fitting or similar work, an agreed proportion is held back for a six months' 'defect liability period'. Design fees are regulated (in a purely advisory way) by the CSD, whose publications should be consulted, or for architects by the RIBA; see parts 18 and 21.

24 Inspection at completion stage, and subsequently after any defects are corrected.

25 Final settlement of finance and records. Photography. Filing and storing of records. Job over.

NB It is rash to assume that the final relieved handshake is 'terminal' to a job – as newcomers will soon discover.

The above account is too tidy and logical, and has some rather alarming simplifications. In practice, constant new factors upset the brief, confusing or enriching it, and every kind of contingency will disturb a logical sequence. The design process is anyway a continuous interplay of creative thinking with reference data. However, in jobs of any substantial scale something like this apparent sequence is the only way to uncover the creative possibilities and to realize them. It will be seen that 'communication' becomes of first importance, with every new situation making its own quite distinct demands.

Here, we shall discuss the ways and means of sending and receiving and getting information relevant to a purpose in hand, that purpose being the whole conduct of the design process as it affects a designer, his colleagues, the clients for whom he works, the contractor or manufacturer, the user or public, and all other parties or interests involved. This is a laboured definition, but it must be clear that here is purposeful communication used as a means – not an end – in situations of which the outlines are fairly well known. Graphic designers can find this confusing, because their 'end' may be to find ways of transmitting a given message effectively. Thus they are using, in the design process, two lines of communication that may either cross or reinforce each other. Other students will be confused by the general tendency of design school drawings, reports, etc, to become in themselves an end rather than a means. It is very necessary to remind students that they will have formed an optimistic impression of the time normally available for design decisions, whether in a design office or in a hard-pressed workshop.

The most cursory study of the events that configure the 'design process' will reveal that effective communication is of their nature. Very many forms and procedures are involved whereby clients are questioned, research is carried out, information is sifted and stored, matters are discussed and intentions aired or confirmed, designs are initiated and developed, and action made appropriate by laborious processes of instruction.

An experienced designer will call upon experience; but will find it useful to ask certain questions as he prepares a drawing, writes a letter, or considers the scale and purpose of a model. The following notes place such questions in their context and suggest something of their nature; though until the evidence has been experienced or at least examined in some detail, it is wise to approach any procedural suggestions with some caution. It is not a formula that is required; merely, at this stage, a few guiding considerations.

Looking to the 'job sequence' as evidence, it will be clear that the designer 'communicates' with himself – he exteriorizes his own thinking in drawings or, say, in using a report to think with – and with his colleagues in ways that may be highly personal, yet will have to assume recognizable conventions to the extent that some external response, or feedback, is called for. As the work moves away from him so must his chosen means become precise, until, finally, his instructions to a contractor must be entirely free of ambiguity. Yet a survey drawing, done at the beginning of a job, must be equally matter-of-fact. Again, there is no hard-and-fast distinction between words and drawings, and even while setting out a survey drawing a designer will be thinking in words about the drawing itself and the physical situation of which the drawing is an abstracted analogue. Is it possible, then to do more than feel one's way into an appropriate course of action?

To answer this question, it is necessary to see what is typically achieved by communication methods at every stage of the design process, and then to decide which procedures to use, and how any alternatives might affect the outcome.

Design methodology can usefully borrow from medicine a distinction between diagnostic and prescriptive procedures, though only a design consultant operates in so Olympian a spirit; the ordinary designer goes on to administer the medicine and thus combines the functions of nurse, chemist, and almoner in ensuring his patient's full recovery. Anyway, the idea of diagnosis and prescription is helpful. Broadly speaking, diagnostic work involves the classical apparatus of problem analysis, in greater or less degree according to the complexity of the job, and to the extent that the job does, in fact, constitute a 'problem'. (It is a mistake to suppose that all design jobs are best understood as problem-solving.) In the context we have used as an illustration, there will be three stages of diagnosis:

1   finding out: to observe, measure, assess, question and record – to get the facts and sense impressions which enable

2   sorting: to compare, distinguish, relate, and order the phenomena with which the designer is confronted, to 'unscramble' the mix and to find sets or categories which helpfully accomplish this

3   interpreting: to evaluate the situation thus exposed, to examine its potential in terms of means, effort, scale, kind, and so forth; to adduce the principles and the alternatives on which a prescriptive solution might be based.

Anyone who credits these procedures with 'objectivity' should realize that human judgement will colour and preselect every seemingly objective assessment other than measurable fact reducible to number (and even here strange things can happen). Problem analysis would be a dull and fruitless affair if this were not so. All the more reason, then, for scruple, and for technique. Returning to the analogy with medicine, signs are more easily detected than symptoms. A designer who approaches his task with the singlemindedness of an adding machine must expect a barren diagnosis and often a misleading one. Imagination and empathy are as necessary as keen observation; to use a different vocabulary, the designer has to put himself out to meet the new situation with all his faculties as a human being, not merely shelter behind the privilege of his role-function (shielded by what is expected of him).

Technique is simply 'taking account of the other' in a realistic and effective way; since design is (at best) a socially transactional art (at worst just a transaction) this is necessary. For most people, and in most situations, there is no tidy linear sequence leading to an optimal design 'solution'. If this were the case, then all tidy-minded people would be, in principle, gifted designers. Often a diagrammatic answer may be available by these means – an answer that just covers the facts with a little to spare – and this is not to be despised. At least the client's existence and situation have been recognized. The quality of a design will grow out of the acuity and richness of response that a designer can bring to the opportunities confronting him, and this is a very personal matter; yet, it is also supra-personal in depending upon a relationship in depth between the designer and the full terms of reference for the work in hand. Technique helps all this to happen, not only by unfolding the possibilities and bringing them to light, but by involving the designer's enthusiasm in the process. Spirit and attitude are worth the larger part of logical method, in design practice as in design education.

It remains true that any design hypothesis will require a test-bed for its viability; misdirected effort must be minimized in this, because if there is one thing that design students are sold short on, it is an awareness of the time available for all these things. There is also a tendency for drawings to acquire the false status of end-products (this is worth repeating) whereas a practising designer's drawings are simply means to a given end, with a time limit for their completion.

Because the purpose of any given communication will not be terminal, but part of a network of interchanges of extremely various nature, the occasion or context must never be left out of account; it is not sufficient to think of a simple loop diagram with feedback. Information circuitry is as useful, and as misleading, as conceptual models of the design process in other terms: the dangers are inhibited response, a false sense of 'objectivity' in procedures, rigidly estimated criteria, and a diminished sense of reality in a designer's informal relation to his work. Although such hazards can be described with some subtlety in cybernetic terms, there is no evidence that an acquaintance with information theory helps to bring home the bacon, in the range of opportunities discussed in this book – and some evidence that reliance upon it will produce a bag of bones. Students must unravel this matter for themselves.

The suggestions that follow are a simple way of selecting tools (and sharpening them) in the workshop of the mind. The use of four words is required:

mode: way in which a thing is done, class of thinking or intention exhibited – might be persuasive or informative, analytical or descriptive, interpretive, literal, final, provisional, open or closed as regards conventions, etc.

vehicle: known form or apparatus of communication, recognizable by conventions – might be a perspective drawing, a diagram, a report, a questionnaire, etc.

medium: physical means employed – might be card, cartridge paper, graph paper, wire, perspex, etc.

agent: instrumental means, tool or process – might be pen, typewriter, xerox, tape, duplicating machine, etc.

The following short statement puts the matter in a nutshell and suggests a simple way of evaluating any communication opportunity involving sender and receiver.

Design for designing
: the ins and outs of effective communication

Communication for designers:

is	purposeful (leads to or qualifies action in a design context)
is not	random, speculative, a work of art, self-contained, self-justifying
though	creativity may be so-prompted, so-ordered, and so-experienced
and	this will be obvious in communicating with yourself;
does	therefore, normally presuppose a knowable context, sender, receiver
for which	'context' implies interaction (more than two-way occurrences);
should	use and transmit energy economically,
and	in the forms (vehicles, conventions) that are used
must therefore be	clear, full, simple, direct, necessary, and acceptable (check)
unless	your purpose is better effected otherwise

NB   an 'acceptable' document employs the right tone of voice for receiver and purpose (this goes for physical character – layout, choice of materials, etc – as much as syntax and choice of words)

for a drawing to be acceptable, its purpose must be estimated with precision, and the drawing must use recognizable 'conventions' to optimum effect. Models may be illustrative or analytical, again according to purpose.

ask yourself   *why* this document or drawing or model?

for whom and to whom and from whom?

in *what* time and place and occasion?

ie   *for* what purpose, with what intention, with what (precise) aim?

or for which of several purposes seen to be related?

then   *when* should it be done and when conveyed (may be separate)

how it should be done (less a question than an answer)

thus   why – who – what – which – when – *how*

embrace   context, relevance, completeness, sequence, tone, consistency, usage,

and other things you may wish to consider

thus giving   *relevant form to what you do*

from the possibilities of mode, vehicle, medium, agent, intention

each of which can usefully be distinguished :

they are not the same nor better described as dualities (medium : message)

NB   no statement of this kind can be value-free.

These are the shallows of communication : hence the *apparent* case for navigation by precept.

13 Simple graphics: a strategy

Graphics is purposeful mark-making to convey messages serially (shortening a definition by Richard Hollis). In any culture that allows of conscious form-giving within manufacture or the built environment, there will be a graphic component which will exhibit all or some of the implicit suppositions of that culture (its 'mental climate'). There are no essential differences of principle that distinguish graphic design from design of any other kind; merely a respect for the purpose and the nature of the enterprise. Certain graphic pursuits, such as typography or cartography, are most specialized studies, and to do good work in these areas does demand far more than a sympathetic overview of their aim and attainments. The same might be said, however, for any specialization. In nearly all adjacent areas of design work, designers must be able to understand each other and often to work quite closely together. This will require a superficial knowledge of the relevant discipline, and of its special terminology, but a genuine understanding will spring from a cultural sensibility held and exercised in common. The fact that this is often lacking – as, for instance, by some architects in their relations with graphic or other designers – reflects partly the divisions in design education, and other factors such as the splitting-off of so much graphic talent into the special world of high-pressure advertising. The distinctions here are not (in other words) to be seen as intrinsic, so much as circumstantially determined. Nor is advertising uniquely destructive to social confidence and the possibility of decent and truthful exchanges between people; high-rise building has recently made architects as publicly 'disreputable' as these willing hyenas of the graphic trades.

This becomes, no doubt, an arguable matter. However, graphic opportunities occur in simpler case: namely, in the day-to-day content of a designer's work as described in the reference section of this book. In this respect all designers, including architects, are practising graphics for the larger part of their working lives; a fact that is commonly overlooked. Drawings, diagrams, schedules, specifications, reports, letters – all these demand graphic decisions, which can be positive in character, or merely allowed to happen. This 'designing design' activity can have a useful effect upon the work thus served; in fact the two (the main design job, and the means for carrying it out) interpenetrate in a most interesting way. The ugly word 'communication' is helpful because it suggests an intention or desired effect (which the word 'graphics' fails to do) and because it broadens the base to include (for instance) oral questioning and model-making, neither of which involve mark-making as such. There is some confusion of terminology in all these matters. Semiotics offers a set of logical relations, and a vocabulary, that seems at first sight a useful clarification; but in practice turns out to be so dauntingly sterile and pretentious that few have the patience to grapple with it. Keeping theory to a minimum, what might be a useful strategy for beginning in graphics, and perhaps especially for architects and three-dimensional designers who need to study the subject at an elementary level?

To recapitulate: designers other than graphic specialists should see their graphic work in the simplest terms, if for nothing else as a relevantly pleasurable way of making marks on paper. They can find a practical usefulness by setting this work firmly in the full context of communication procedure in the design process. Notes on context, and criteria for evaluation, are given in part 12. Good graphic design requires an effective relation between *alphabet, image, medium*, and *process*. In a practical way, it is probably best to advance on all fronts at once:

1   Examine every graphic opportunity that normally occurs from the moment a designer accepts a commission to the ultimate filing-away of job records. Consider what has to be done to make this work enjoyable and effective.

2   Assemble a collection of useful information, including typesheets, paper samples, etc, headed by an address list and a source-list for further information when you need it, and not forgetting a key-word graphics terminology – Ken Garland's *Illustrated graphics glossary* will help.

3   Investigate in outline the available printing processes and what is required of a layout or artwork for such purposes.

4   It is good to experiment every time you make a mark on paper, by hand or machine: why, how, when, where, what? More is learned from experiment than from perception theory, but simple logic or set

theory may prove suggestive when considering the typical problems of tabular setting or organizing complex messages.

5   Consider what you have to deal with. If we read from top to bottom and from left to right, is a page or a rectangle a 'neutral' field? – and if not, what are the consequences for sequence and layout? How are units of meaning formed most effectively out of alphabet? If a paragraph is one such, think of half a dozen ways of forming a paragraph – and what is entailed. When alphabet is developed into syntax, the prime ingredients of order are *distinction, emphasis,* and *grouping,* from which a clear graphic *sequence* is derived. How might such a realization affect your attitude to scale, structure, the form of a report, the order of priorities (and their recognition) in a report? There are many considerations here and most of them can be thought through without specialized knowledge.

6   Come to terms with the graphic potential of a portable typewriter. The elephantine desk machine is useless for layout work and very discouraging to beginners. Touch-typing is a skill worth having but its possession may entail bad graphic habits: some training courses are notorious in this respect. Two-finger typing is still faster than handwriting (with a bit of practice) so for layout purposes and occasional use, this is an acceptable alternative to the proper thing. The typewriter is good for controlled experimentation because there are mechanical constraints to be respected (for a definition of design as converting constraints into opportunities, see many references elsewhere in this book). Such constraints are a pleasure to explore (and will not be listed here) but they include, of course, a built-in asymmetry deriving from the 'ranged-left and open-ended' nature of the platen movement (usually ignored by typing instructors). It is possible to study very profitably all the organizational disciplines of information graphics (alphabet) and to follow through the implications of 5 above. There are then the various reprographic alternatives, from the old-fashioned duplicator through xerox to the more ambitious possibilities of offset litho printing.

7   Finally, other people's work can be examined; at first hand, or, less satisfactorily, in reproduction – including of course people like Lissitsky and Max Bill who bridge across from graphics into other areas of design.

Designers who want to specialize, to become 'graphic designers' as such, will not find themselves lacking in advice or instruction from the very many schools. Such students might also wish to ask themselves just what order of reality is embraced by this work, what role-seeking and role-making is implied, how far graphic design can be a responsible trade when after all, most messages conveyed are not of the designer's choosing, and lastly, what areas of graphic design suit their own predilections and offer some experience of authenticity.

NB The suggestions in this part will find reinforcement in parts 12 ('Communication for designers'), 14 ('Drawings and models') and 17 ('Reports and report writing'). Students in schools are always obliged to document their work towards assessments and their final degree or diploma. This is a special variant on the office job-file or job-record. It is worth doing the thing properly by making an intelligible graphic sequence. The design problems are always interesting and much will be learned in the way of concise copy-writing, co-ordinates for paper sizes including folds, and consistent means of identification and retrieval. The following may be helpful as a possible check-list for such a task (for a larger job in any area of design):

1   Legend (name, job title, job number, date(s), etc.)
2   List of contents
3   Briefs (initial, and revised)
4   Note of academic situation or objectives if relevant
5   Summary of relevant job facts and other data
6   Analysis of requirements
7   Evaluation of possibilities latent in the job
8   Proposals in principle (alternatives)
9   Final proposal and reasons
10 Notes of materials, fittings, manufacture
11 Notes *re* costing
12 Report (which would normally include items 5-11 or even 1-11)
13 Action taken, modifications
14 Presentation drawings, notes, photograph of model
15 Copies of working drawings, specification, schedules
16 Machining or other instructions
17 Photographs of job process, assembly, other models, etc.
18 Photographs of final outcome
19 Test comments or user reactions generally
20 Correspondence and other job documents
21 Afterthoughts (your own criticism)
22 Job duration record (times, dates)

14   Drawings and models

It is very important to distinguish one kind of drawing or model from another – by never failing to consider its purpose, occasion, and recipient, and therefore its nature. Confusion would be avoided, and tedium minimized, if draughtsmanship were *always* studied in the general context of communication theory and practice; but this does not frequently happen.

Fast and skilful technical drawing as carried out in a design office is just a matter of long practice: students should not be discouraged by their first indistinct attempts. I am speaking solely of competence in technical drawing; not of satisfaction. There appear to be no short cuts to this technical facility but common sense and a clear definition of a drawing's purpose will give a head start. Free-hand drawing is something of a personal gift, valuable to a designer but not as essential to every kind of work as is usually thought – a designer can design well and not be able to draw in this way at all; conversely, a designer who draws marvellously may be mistaking his vocation as an illustrator or painter. Much design drawing – of an informal kind – is no more technically 'gifted' than mental sums transferred on to paper. Designers think and talk with sketches and diagrams, sections, full-size details; a fluent and personal use of diagrammatic technique is certainly necessary. Such work can be better seen as analytical mark-making than as analytical drawing in the classical sense. Observational drawing helps focus observation, and has all the well-attested and richer benefits from a co-ordination of hand and eye, but the transference of such gains into design procedure is an indirect one – as would be from comparable activity of other kinds.

Nothing should be taken for granted in the way of materials and instruments. The qualitative differences must be explored, compared, contrasted; if only because much of the pleasure of drawing (or modelmaking) comes from a subtle rightness in the relation of tool or instrument to material. Anyone who uses blunt tools, or confuses a Graphos pen with a clutch pencil, or who treats paper as a neutral non-material, will miss the benefit of good technique as a feedback into design thinking. These are fine-tolerance perceptions carried out at the finger tips, but awareness of them can help the tone of design responses in a very encouraging way.

Drawing practice and modelmaking will occupy a large part of a designer's time, and in a design school, even more. There will always be formal instruction in schools, but the following simple distinctions will help to place these skills in a communication context:

Diagrams are abstract, partial, energetic, concerned to establish or convey ideas and values directly, thus having an analytical or interpretive purpose. Usually have open conventions (excepting graphs and mathematical conventions), may be imprecise, or may be examining exact quantities, usually have diagnostic function.

Illustrations (including thumb-nail sketches and 'perspectives') are depictive, present appearances from which inferences may be drawn, are often atmospheric in nature and persuasive in purpose, have closed conventions. Usually have prescriptive function; better for presenting conclusions than determining them.

Surveys are records of measured and verifiable fact reduced to quantities, though survey drawings may be accompanied by interpretive notes. Closed conventions. Diagnostic function.

Working drawings are strictly purposeful and are instructions. Use rigid but propulsive conventions (i.e. lead to required action). Many types according to purpose, occasion, and recipient. Prescriptive function.

Words often have a function supplementary to drawings and models: students should note the importance of this. Many kinds of drawing and most models are too abstracted to be fully indicative of their intention. Words in this context may enquire, annotate, argue, describe, record, and otherwise supplement drawings and models, and may be written or oral or occasionally recorded on tape.

Signs, numbers, codes, etc. – clarify, regulate, supplement.

Such distinctions are helpful because they indicate the design value of the choices available – the correspondence between their form and their meaning. Once this is grasped, the whole subject ceases to be dead and conventionalized and becomes purposefully allied to other communication procedures. The worst handicap to a designer is the unintelligent way in which 'technical drawing' is sometimes taught, as a self-contained discipline quite outside any purposeful context. The point is worth repeating because some designers have to relearn technical skills and they should know that an apparent asset may turn out to be a communication liability.

The main thing is to know when a drawing can be 'free' and open in its conventions, or when it has to meet well-known and agreed conventions as are, for instance, specified in the British Standard recommendations for building drawing practice (BS 1192). As a safe general rule: *always give too much information* rather than too little. Drawings are no more than messages and records; they must convey fully what you intend, with no room for misunderstanding. It is a mistake to worry about the immaculate appearance of a drawing when it must be *for use*. In design schools, drawings become 'exhibits'. Use a notes column, on the right of the drawing.

When considering how to do a drawing, and what kind it should be, refer to the check-list at the end of part 12, and never forget to ask yourself, who will use this drawing, for what purpose, and in what context?

The distinctions of *kind* in drawings and models given here, are very basic: with experience, much refinement and experiment is possible, as in the development of different kinds of component and assembly drawings. This may arise either from the known context of use, or from the nature of a particular design. An example might be joinery detailing drawn out for a contractor already well-known to the designer, in which numbered components are presented as in a cutting list for the machinist.

Models are always encouraging to work with and can be a valid form of presentation. The subject profits from a very thorough exploration of the technical possibilities, on the one hand, and of the purposes involved, on the other (again, see end of part 12). Full-scale mock-ups of joint details are sometimes helpful, full-scale but drastically simplified environments are occasionally worthwhile, and of course furniture requires model-making at all stages, to the point where prototypes are under test. Scale models range from the quick fold-up diagram to its opposite, the doll's house illustration (see categories above) and between these extremes there is much scope for experiment.

A designer thinking out a job will go backwards and forwards from one kind of drawing or model to another; testing and exploring. Thus, as was said of communication in general, there will be a balance between subjective freedom and objective scruple; as the job develops, 'the chosen means become precise'.

15   Survey before plan

As when Le Corbusier says that 'architecture is organization', the
notion of survey is shorthand rhetoric; implying that very much more
is connoted, or at least hinted at, than the simple requirement of a
clear and adequate drawing – of what? – the ground plan? What is
implied, centrally, is the significance of the plan (both the intention,
the fore-cast, and the horizontal section) and of its *grounded and
verifiable base* in number (reduction to quantities) and fact (as distinct
from opinion or judgement). Another banner unfurls: 'the plan is
the generator'. Thus the notion of 'survey' acquires, if not mystical
overtones, then certainly a special weight of responsibility. If realism
is the watchword, don't con yourself or anyone else; get the facts
right and get all the facts. Test every step forward. The attraction of
such polemic is almost a measure of its essential fallibility. Even at
the simplest executive level, that of measurement and record, for
which prior status is claimed by the seeming neutrality of the
procedures, we know that human beings are impossibly inventive,
both in the selectivity of their vision, and the wild unreliability of
their memories. With respect to their deeper fallibility we are now
as chary of Five Year Plans as of planners in general, with their
sometimes compulsive itch for tidiness at the cost of spontaneity; yet
the notion of survey, of finding out what you can *first* and as
accurately as you can, is still intactly a reasonable undertaking
wherever quantities are in question.

The notes that follow are an aid to so doing, at the simple level of
a space or a building that requires to be measured; free of all cosmic
misgiving as to what might be done with it. Readers must form their
own weighting of such implicitness. Designers will be aware that a
surveyor is a specialist, one upon whose services a designer may have
to lean occasionally, and it should be clear that these introductory
notes do not approach this level of expertise, nor do they involve
equipment for which special tuition is necessary. Even so, it is
surprising how the simplest job (say, measuring up a room or a flat
or a small shop with a view to conversion or fitting out) can go
wrong, prove to be wrongly or insufficiently observed and recorded.

The context of such a task goes in with the designer in two respects. First, in a literal way the surrounding context of any such small volume will need to be looked into, analysed, and in certain ways questioned and recorded. Second, the character of the spaces involved will not only commend themselves (as having certain features, relations, categories) but if at this stage the job itself is known, or is known to be 'one of a kind' already familiar to the designer, it is inevitable that matching will begin to take place. This matching will tend to produce 'fit' between known categories of use or function in the job, or categories that are typical of it, to differentiation within the spaces that are being surveyed. In other words, ideas jump the gun and begin to form around the special formal attributes that begin to be revealed. This occurs as the structural logic of a building begins to unfold, so that 'surfaces' begin to have distinctive character instead of being undifferentiated (for example, as between the outward expression of a load-bearing wall built of brick, and that of a partition wall built of plaster and laths on studding). Gradually the essential structure and volume stand revealed together with accessory situations that may well be swept away or modified.

These matters are not easy to describe; the real point here is that these survey procedures (like asking questions, writing reports, etc.) are *already design* and should be approached in that spirit. A survey should never be treated as a technical exercise. The imagination should be at work discovering such spatial possibilities as seem inherently available, and matching them with imagined uses. This is valid and natural (not 'formalist') because the designer will include within his ordinary competence an adequate testing of every hypothesis he forms, upon the test-bed of 'requiredness' (a Gestalt term) in the job as this in turn begins to assume a structure. The inherent logical relations of the one (the building) must work for those of the other (the job, or 'the problem' or 'the opportunities revealed'). For these reasons attitude of mind is very important in survey work (as in diagnostic procedures generally). Alertness is a state of mind that can be summoned and sustained by an act of will. A new situation must be entered watchfully, but also with a certain leniency of predisposition, a willingness to be engaged (to enter into transactional parlance). As when asking questions a designer must learn how to listen, so he must learn to absorb impressions in survey work; a building must be felt for. Given experience, technique, and a productive attitude that always has an ingredient of respect, a building or spatial situation will be found to speak for itself. Not, however, if a designer has never learned to listen. Although most remarks in this part (and all those that follow) are about building surveys, there is a general sense in which any situation of 'survey' will yield responsively to the right balance of sympathy and disciplined enquiry, of which careful measurement and drawing are merely component experiences. It follows of course that if a designer wishes to *exclude* personal response from a survey, it is better to have the job done by a professional surveyor (or colleague).

Getting down to brass tacks, and taking the simple building survey as the typical case for the rest of this part, the notes begin with some guiding considerations, and there follows a check-list that should be of practical help.

1   Do your homework first by gathering preliminary information including district maps.

2   Make sure you have a full kit of instruments as described in the check-list; if not, go back for anything missing *before* making a start.

3   Be prepared to note everything down; but *everything*. The commonest mistake by beginners is trusting to memory. It is more realistic to suppose that nothing exists of which you have no record.

4   Use overlays to separate out different kinds of information (formed simply by using thin detail or tracing paper) and have two underlay grids to help free-hand drawing (squared and iso- or axonometric). See check-list. It is essential to have a mounting board, with clip, so that sheets of notes are under control rather than on a dusty floor. Supplement all measurement with the camera and sketching.

5   Enter old buildings alertly, on the prowl for trouble. Note any evidence of smell, subsidence, cracking, rot, woodworm, damp, loose plaster, stuck doors, pattern staining, damaged fittings.

6   Note subjective impressions immediately (actually note them on paper).

7   Number, write, and draw clearly, even if it seems slow; if not you will always regret it (and end up wasting more time).

8   The same applies to the thoroughness with which you take and note measurements. It wastes much more time to rush the job and to be obliged to come back (perhaps more than once) for missing information.

9   Always note the superficial nature and conditions of surfaces, but

10   Always go beyond surfaces, to structure, and to an awareness of materials. When you have to make intelligent guesses, say so.

11   Always note and differentiate load-bearing walls from (potentially movable) partition walls and from party walls (common to adjacent premises) and measure apertures in their true structural sizes (linings can be noted separately).

12   A careful record of joist intervals is essential, future load-bearing decisions may depend on it. Also ceiling rafters, studding, and any other concealed relevant information that could affect fixtures and fittings.

13   Measure from one point, line, or surface, or errors will be cumulative. Work off struck chalk datum lines, off which measurements can be triangulated at regular intervals. Always use diagonals, and plenty of them.

14   Make full-size drawings of mouldings, architraves, etc, using when necessary a pin moulding template.

15   Note vertical heights that may be worth respecting (apart from those noted as a matter of course).

16   Estimate and carefully note all relevant information to extend the space you are concerned with into its immediate environment. This will include, obviously, 'the neighbours', but also physical situations like drainage, structural matters, facade or elevational relationships, light, views, aspect, and a service analysis to include waste disposal, plumbing, electricity, deliveries, etc. (see check-list).

17   Note any matters of possible concern to by-laws and local authorities (this needs experience).

18   Note all socket outlets, plumbing, etc, but do not waste time carefully drawing up superficial fittings that are obviously expendable.

Survey check-list
excluding special instruments

take   Container (purpose-made)
Carrier board with clip, hardboard or ply, size to suit (min: 25 x 40 cm)
Paper, detail and/or tracing
Grid underlays
Pencils and pens, include colours
Eraser
Penknife and/or small pencil sharpener
Chalk and lines
Drawing pins and lead weights
Sellotape
Plumbline
Pocket rule-tape (imperial and metric), 3 metres
Rule-tape not less than 10 metres
Folding boxwood rule, 2 metres, for verticals
Pocket compass
Spirit level
Water levels if required
Angle bevel and protractor
Pin moulding template
Short metal rule (for carrier board)
Torch, batteries checked
Probe or birdcage awl
Camera with spare film and flash
Hammer (light)
Tape-recorder if relevant
Maps, drawings, case notes, keys, relevant phone numbers
Small stoppered container for specimens

NB   the above will fit into a very small container made for the job;
survey can be done (and often is) with *less* than this elementary
equipment, but there are sensible reasons for all inclusions

check   Area notes by map and observation
Street access, width, pavements, planting, drainage
Traffic flow, direction, feeder sources, volume
Traffic parking (question for after working hours)
Bus services and stops (and underground if relevant)
Train services and station access
Noise

Litter
Street aspect by compass
Relevant heights and light cut-off
Stop-cocks and other control points for services
Waste collection arrangements
Delivery arrangements
Telephone
Line TV and aerials
Water, gas and electricity supplies
Drainage
Gardens and planting
Facilities: doctor, hospital, library, laundry, pub, stores
Milk deliveries
Garaging
Pram or wheeled access
Adjoining buildings
Other possible interests
Roof condition and structures (as estimated without ladders)
Exterior wall condition
Pointing, gables, barge-boards, soffits, trim, footings, steps, paths
Exterior doors and windows
Locks
Provisions for post and deliveries
Bells, handles, porches, mats
Security measures

First impressions of interior and entry
Hall or reception area
Stairs
Lifts
Sound
Light
Smell
Temperature
Humidity
Maintenance
Surfaces, paintwork, paper, plaster, etc.
Structure
Materials
Circulation
Doors
Skirting

Floors and floorboards (widths, direction, material, pattern)
Joists, trimmers
Hearths
Fireplaces
Vents
Windows
Ceilings
Rafters
Mouldings
Built-in cupboards
Light
Aspect
Views
Electricity gas and water switches, taps, outlets, etc.
Telephone and TV
Metering for services
Heating systems
Insulation
Washing and bathing facilities
Kitchen and serving facilities
Cleaning facilities
WC facilities
Attic access
Roof structure and condition
Tanks and other services
Guttering and downpipes

NB   estimate age and condition wherever relevant in a survey (and note)

16   Asking questions

The notes are confined to enquiry in the particulars of design practice. More general and more searching questions are professionally relevant – to fail to ask them is to invite creative anaemia – but such questioning is ill-served by precept. Though perhaps it helps to jump over mole-hills before climbing mountains.

Most design problems will be presented in ways that may be diffuse, ill-defined, or actually misleading; yet, a problem cannot be resolved in any satisfactory sense before its nature has been determined. According to the 'problem', such discovery may be instantaneous, or the problem itself simply invented by the designer in an extremely open situation; in either of these cases it is unhelpful to use the word, or to conceive of design as problem-solving: a different order of resourcefulness will be required. However, keeping to the work which has been used as an illustration, the designer will be asking himself and others a continuous stream of questions, some of which may need to be formally stated or written down.

The reasons for this:

1   To gain an awareness of the problem that is sufficiently objective, i.e. takes full account of relevant facts, relevant interests, relevant possibilities, relevant limitations.

2   To focus and clarify your own response to the job and to exclude (as far as possible) irrelevant responses – those that might belong to different jobs still strong in memory, that might be a simple projection of your own interests, or that might prejudge certain issues before you have investigated them.

3   To enable you to sense out the feel, weight, context of the job and relate yourself productively to its potential (to turn yourself toward it).

4   To enable you to construct a satisfactory working brief, i.e. clear and agreed terms of reference.

General considerations:

1   The extent to which formal questioning is necessary will vary from job to job, but has nothing to do with the *apparent* scale or simplicity of the problem you encounter – questioning may extend or alter the possibilities beyond recognition.

2   It should be recognized that diagnostic work has a creative component, a kind of dialogue between you and the client or between you and the total situation you are examining; thus its instrumental success very much depends on your own attitude and approach. To ask the right questions is not to carry out a tiresome preliminary to design; it is already design, and it may require considerable imaginative effort on your part. In the answer to one question is the genesis of the next.

3   Questions served up 'cold' will harvest facts, but may leave untouched all manner of subjective or non-measurable factors which may be crucial to the understanding of a problem. Most (not all) problems will involve a client; most will involve human relationships, and therefore tact and an effort of identification with other people's viewpoints.

4   Your intention in asking questions must be reflected both in their structure and in their tone; categories of question useful to you may not be useful to the client or invoke helpful associations in his mind – only later, as abstract categories assume concrete reference, will you be talking on convergent terms with your client.

5   (Tone: 'the speaker . . . chooses or arranges his words differently as his audience varies, in automatic or deliberate recognition of his relation to them. The tone of his utterance reflects his awareness of this relation, his sense of how he stands towards those he is addressing' – I.A.Richards). A suitable tone does not always imply informality, or prevent you from arranging your questions into categories, but may well require a careful choice of terms in describing them or introducing them.

6   The designer is acting as an agent in a process which (viewed in retrospect) may appear to have discovered itself over an extended period of time; questioning takes place across the flux of decision-making in a variety of situations. Early on, personal observation will involve self-questioning.

7   There is often value in mixing questions by free association; but concealed or subsequent categorizing will in itself provide association, and categories relevant to questions will usually be relevant to the answers.

8   Many questions are best put in a form that suggests typical answers, or alternatively put as assumptions that can be agreed with or corrected. Such assumptions need framing with some care, or you may cut out their possible suggestiveness and thus 'short circuit' back to where you started. Stated assumptions work best at the extremes – either as a mild confirmation of something fairly obvious, or as something so outrageous that a lively response is inevitable.

9   It is a mistake to rely on formal means to establish informal truths. Facts are best investigated formally; opinions or attitudes informally. A client may shy away from an attitude imputed to him by the way you have put a question in writing, whereas face-to-face things would have been different.

10 In formulating written questions, remember it is difficult for someone to withdraw a direct answer to which he has committed himself, unless you can show reason that he was misinformed. Commitment may be useful on matters of fact, but on matters of opinion may merely express prejudice. Taking this further, if you are working with someone toward some common end, you must allow their prejudice as much breathing space as your own; but don't carelessly drive them into positions from which they can't retract without loss of face, and which may come to impinge on your common activity quite needlessly.

11 Of course facts and opinions will be inextricably mixed in your client's experience of his problem; hence the delicacy of your diagnostic technique.

12 As a rider to that, be careful not to confuse diagnosis with cure. Questions seek answers; not a euphoria of perpetual doubt.

13 Loosely and vaguely-put written questions are quite useless. A 'portmanteau' question will receive a portmanteau answer, and also diminish your client's confidence ('what is he getting at?'). Questions are best put singly and profusely – unless you have a particular purpose in mind.

14 Consider carefully when and when not to use tapes, cameras, and notebooks. The contextual requirement (in the design sequence, see part 11) must be balanced against consideration of tact and judgement.

15 Questions and answers qualify and reinforce – *but do not substitute* – judgement and decision.

Oral questioning at informal meetings:

1 Above all, remember that to question is not to interrogate.

2 Mental alertness is essential (not so obvious).

3 Leading questions may be fruitful as guides to intention or attitude in those present; questions can be put in this way at such meetings that could not be asked in more formal circumstances (or in writing).

4 Do not take your answers too literally – unless they refer to matters of fact (which can always be checked) it is necessary to take conversational answers or assertions at less than their face value.

5 It is difficult to establish needs by direct questioning, but relatively easy to establish desires, preferences, prejudice, and opinion based on specific experiences. Motivation is far more elusive.

6 Do not allow the nature or sequence of questions to get on to 'tramlines' – closed patterns of thinking – which may give you a false run of evidence.

7 Avoid giving an impression (which should anyway be unwarranted) that you have made up your mind and are merely asking questions as a matter of form.

8 Make some allowance for positive and negative appearing in a misleading juxtaposition: e.g. if challenged, a person with authoritarian habits may stress his anti-authoritarian ideas or desires. Most people have inbuilt compensations for exaggerated personality traits and they are delighted to be able to exercise them.

9 It is off-putting to ask questions conversationally from your notebook.

10 A useful extension of questioning technique is the meeting at which all parties concerned air their views at random, with you keeping the thing going as a sort of group therapist. The purpose of such a meeting should be understood by all concerned. Avoid the meeting becoming a collective instructional session leaving no subsequent room for manoeuvre. Such meetings can be time-savers, particularly where many people are involved, but they should be followed up by more scrupulous enquiry, whether or not such enquiry starts from inferences drawn from the meeting.

NB In all this the designer is meeting a situation, not attempting to master it. The only value of conscious technique here is in making the designer more useful, because more accurately responsive, to the situation he is serving, and perhaps a bit less liable to take his own unquestioning assumptions for granted.

Reminder:

During the diagnostic phase of design work you will (with respect to what confronts you and what you have to seek out) observe, compare, relate, distinguish, question, discuss, 'research', measure, estimate, and record. This might be said to be your diagnostic shopping list.

When considering the constraints in a job (constraints: 'a restricting condition' [Chambers], factors that impinge upon a job, that must be taken into account, that belong to its terms of reference but may be unstated as such) – remember two things. First, a necessary distinction between constants and variables, and second, that you must search for *agreement, conflict, latitude, and determination* within your constraint system.

Finally, it is quite proper to redefine your intention and objectives in the light of a constraint and opportunity analysis, and from there to establish the principles that seem necessary to respect both; but a design *concept* cannot be 'defined' as an embodiment of principles; it merely has a structure of required argument to be offered against. Otherwise design would indeed be very easy.

The constraints on which students are usually (and understandably) weak, are the budget, restrictions in law, the allowable time for design work in its various aspects, and design predelictions and/or experience (which will slant an individual designer's approach, however 'objective' he tries to be). It is necessary to know that such factors may be weightily influential in design decisions.

17 Reports and report writing

A report is a written statement of fact and opinion, addressed to clients or to colleagues (or to oneself). A report may be retrospective, and therefore a terminal objective in some series of happenings, but usually its purpose is seen as a more active one: to support, explain, or gain agreement to a suggested course of action.

Designers have good reasons for employing the report as a familiar and often indispensable communication vehicle. Used properly, a report will more than repay the labour of its preparation. A report can describe, interpret, and analyse a situation with sensitivity and suggestiveness, and in language that any ordinary person will understand. For clients, or indeed anyone else whose interests must be consulted, there are no difficulties of transposition (as from drawings to words) that might inhibit a full preliminary discussion. There might of course be transpositive problems of style, if the report has been badly done (failing to respect its purpose and recipient). In the early stages of a job, a client may be wary of a designer's drawings, however seductively presented: the medium is not his own, and may conceal all manner of implied decisions which he can only take on trust. Sometimes the client will discharge his unease by demanding entirely unnecessary alterations, or he may even reject the first design out of hand, until he has been reassured by a set of drawn-out alternatives. This is a very hit-and-miss way of going on. A preliminary report gives confidence, because here is evidence that the problems have been examined with professional thoroughness. There is also the possibility of active intervention on his part *before* the design reaches the drawing board. A client who enjoys a real sense of participation in all the thinking that surrounds the design process may be ready to accept a radical and (for him) unfamiliar design solution, simply because he has understood the genesis of the solution in the problem itself. Thus may a very conservative client find himself embarking upon inconceivable adventures, and enjoying them.

These are obvious benefits to a designer's work, but reports can do far more than this primitive midwifery. In work of complex nature, reports may have the key role of a more subtle, creative, and constituent *process*. The writing of a report will order (and externalize) a designer's own thinking, and will enable him to call in question his own working assumptions – before it is too late. The designer will not only avoid misunderstandings with his client – a negative gain – but may also attain a new understanding of the task, both from the discussion that such a report will inevitably precipitate, and from the effort involved in arguing out a problem analysis. Normally a report is neither a brief nor an analysis as such, but rather a distillation from analytical thinking; expressed in terms which will further a diagnosis into relevant action. A brief must always be redefined when feasibility studies are complete, or when the problem can be freshly seen after due thought and examination by the designer. A report may include a suggested redefinition, or may merely point to its necessity. Finally, it must be realized that a large-scale job is a co-operative venture. It is a courtesy to give all concerned a document that makes plain the reasons for a design and their own part in it (this applies to design colleagues and consultants, but also to contractors).

It is sometimes said that all this is far better done over a glass of sherry at the club (designers are assumed to feel at home in all possible worlds) or perhaps during the main course of one of those interminable businessmen's lunches. In practice, a designer will have any number of informal meetings with his clients, at any one of which the burden of a report may be freely discussed. A report is a formal document that can be taken home and considered at leisure, or may be passed for comment to business associates, and to others whose interests may be involved. A report is also, indubitably, a record (though not a legally binding one). Reports have their own usefulness – they are not a substitute for conversations, contracts, letters, questionnaires, drawings, or other vehicles of communication.

Reports should be reasonably concise and readable (obviously), but beginners sometimes forget that style (meaning here, tone of voice) is equally important. Those who lack experience of writing should use short words and short sentences, making quite sure that the structure of the report has within it a clear and logical sequence. More confident writers will tend to adjust the tone of their writing to their purpose and to their awareness of the reader. At all costs avoid bombast, rhetoric, excessive technicality, and those knowing turns of phrase that imply ignorance on the part of the reader. A report so written may have unfortunate consequences.

The following notes summarize the active functions of a report, and the more obvious factors that should be kept in mind. There are less stringent requirements for the kind of report that merely supplements a set of detailed proposals (e.g. presentation drawings).

Reports:

1   must be written in full awareness of their purpose and occasion and the receiver (the nature, purpose, and occasion of a report should always be stated at the beginning, however informally)

2   require an imaginative awareness of the person(s) addressed, and thus an appropriate *tone* of address

3   should communicate in a direct, sequential, and consistent manner (derived from 1 and 2)

4   have no 'correct' structure, merely an appropriate one

5   will generally follow the logical sequence of given-required-proposed, though not necessarily a *formal* structure

6   should normally keep clear of design jargon and private language, using the word 'I' with discretion

7   should argue syntactically, adding questions or known facts or quantities in tabular form (preferably separated as a supplement to the report)

8   may include sketches, diagrams, graphs, photographs, etc, either as a supplement or directly related to the text

9   *if lengthy, should use short paragraphs which can be numbered* to facilitate reference back

10 *if lengthy, should also begin with a summary of conclusions* but not, desirably, a summary of recommendations as such (better to let the argument require them)

11 may be read (and handled) by several people; to advantage, several (numbered) copies may be offered

12 must be complete and self-explanatory (the writer may not be present to make good any omissions)

13 should provide a professional interpretation of the problem that concerns your client

14 should thus establish for the client the full implications of the brief he has given the designer

15 should relate such implications to the 'constraints' in the problem (e.g. time, money, space, legal restrictions, etc, etc.)

16 should thus provide both an insight into the client's needs and the intentions behind the brief, and an estimate of the true possibilities latent in the brief

17 should thus clarify for designer and client the margin of free choice available

18 should certainly make clear the *context* of present decisions or recommended action

19 unless concerned with specific decisions, will be concerned to adduce relevant *principles* which such action must respect

20 most important of all, should gain a sense of *priority* both in the needs and potential of a situation

21 should seek a client's agreement to a designer's working assumptions, and any forseeable issue of principle that may bias the design approach

22 should recommend and define subsequent action (how, when, what)

23 may anticipate (require, provide for) a positive response from the client or persons addressed

24 may lead to a fresh and more explicit definition of the design brief

25 should be a creative task in the design process (constituent, not accessory)

26 should thus be directly helpful to the designer, and may even in some cases be written to himself or to colleagues concerned with the job

27 should be regarded as a design opportunity – construction, layout, materials, should not be overworked in relation to the content of the report, but should look and feel *right* in terms of 1-4

In summary, it may be said that report writing has three special merits: the conventions are as open as the use of the English language, and can be shaped accordingly (to a close match of form and purpose); a report is a prompt to active participation by all concerned, and is thus both a catalyst and a vehicle of consultation;

and a report can function with salutary effect at different stages of
the design process – notably, those of analysis and presentation
(see parts 11 and 12). A report can also, of course, provide
conclusions, and can thus be (for a consultant's purpose) the required
end-result of his assignment. For some designers – by no means all –
report writing can be a good way of 'talking out' a job, or
considerations that surround it, in a way that is more reflective and
logically sequential than ordinary talk, and less transitory. For all
these reasons, here is a skill worth learning – necessarily, by arduous
practice. Designers for whom writing is not congenial, can use the
report device at its leanest and most purposeful : as a tabular listing
of relevant notes that simply accompany drawings and/or models;
or as tabular notes that accompany spoken presentation of proposals.
(Tabular : like a shopping list.) The check-list at the end of part 12
should be helpful in writing reports of any kind.

18 Finding out for yourself (GB)

NB *addresses and telephone numbers are in part 21*

The following notes review common sources of information for
designers, with a necessary bias towards England and London.
How do you find out about something? Answer: (a) suck it and see,
(b) ask someone, (c) look it up. The notes refine these possibilities.
For simple enquiries on matters of fact, always remember the
telephone. Letters may get more serious attention. Compile your
own list: the suggestions here are obviously selective and open-ended.

1   Telephone directories: ordinary A/Z and Yellow Pages (classified)
for trade. Your main post office and public library will usually keep
directories for the principal cities in the UK.

2   Standard works: a few must be owned if a library isn't handy of
access. Should be known not in detail, but for what you are likely to
find, and where. See part 19 ('Using libraries'); and some examples:

Chambers twentieth century dictionary (among the concise
dictionaries; backed up by the *Shorter Oxford,* usually a library
resource)
Fowler's *Dictionary of modern English usage*
Partridge's *Usage and abusage*

Encyclopaedias (usually a library resource)
Specification (essential for architects and 3D designers)
British Standards Institution *Yearbook* (see 9 below)
Building Centre and CIRIA guide to sources of information
AJ (*Architects' Journal*) information sheets
Directories of associations and professional bodies
Specialized directories and year books

Telephone directories
Local maps
A to Z atlas of London
Transport information

3   Libraries: an essential resource, described in part 19.

4   Bookshops. Local and regional bookshops need the support of
    discriminating patronage and will order any book in print, though may
    not stock it. Locally required 'course-books' may, however, be found
    in quantity. These are among the leading London shops:

Design Centre Bookshop (28 Haymarket, SW1): broad range serving
most design disciplines. Mail-order lists issued.

Dillon's Bookstore (1 Malet Street, WC1): excellent general and
academic stock; with a specialist art, architecture and design branch
elsewhere (8 Long Acre, WC2).

Triangle Bookshop (36 Bedford Square, WC1: below the Architectural
Association): expert personal service for architecture and some
design books, with good foreign-language stock.

Building Centre Bookshop (26 Store Street, WC1): especially good for
technical books (building/construction), with the benefit of the
adjoining exhibition of building materials.

RIBA Bookshop (66 Portland Place, W1): excellent for architectural
and some design books. The library (reference only) is upstairs. The
RIBA issues an annual, classified book list.

Zwemmer's (24 Litchfield Street, WC2): old-established art bookshop
with strong design and architecture sections.

Compendium Bookshop (234 Camden High Street, NW1): large
stocks of alternative or minority interest books and pamphlets.

Freedom Press Bookshop (84b Whitechapel High Street, E1):
old-established small bookshop and publisher, specializes in left-wing
politics and sociology from a libertarian standpoint.

Collet's International Bookshop (129 Charing Cross Road, WC2):
specializes in left-wing politics, from a generally Marxist standpoint.

Housmans Bookshop (5 Caledonian Road, N1): stockist and
distributor of libertarian books, with ecological and peace-movement
emphasis.

(Specifically right-wing interests do not need special mention here:
they are well provided for, through the large retail outlets and the
national media.)

There are many specialist bookshops; for example, two of the best known are:

Stobart (67 Worship Street, EC2): specialist for all aspects of timber technology, and related trades. Lists issued.

Lewis's (136 Gower Street, WC1): medical and scientific books.

5   HMSO bookshops (in London: 49 High Holborn, WC1; and in the main cities): stock a very wide range of government published or sponsored literature. Sectional lists available. Distribute (for example) Building Research Station and Forest Products Research Laboratory publications.

6   Building Centres: of special interest to designers (architects, for example) using constructional materials and methods of all kinds. In most large cities (London: 26 Store Street, WC1). Permanent trade-sponsored exhibitions with information back-up (including catalogues). Host to more specialist trade associations (such as TRADA: see 27).

7   Catalogues: design schools or design offices should have their own technical library, compiled largely from trade catalogues (perhaps classified under CI/SfB – see part 19) and often serviced and provided by the Barbour Index (a commercial technical library system). More general catalogues involving a wide range of products (as from ironmongery wholesalers or printers' suppliers) will become standard reference material. Exhibition catalogues are often useful.

8   Magazines and journals: best known is perhaps *Design* (published by the Design Council). Still staple reading for architects are *Architectural Review* and *Architects' Journal*. There is a fluctuating number of magazines of more specialized interest, among them *Designers' Journal* and *Crafts*; to these might be added the magazines (such as *Blueprint*) that have come with the enlarged public interest in design activity. There are also excellent magazines from abroad. As examples: *Casabella, Domus, Industrial Design,Werk, Form, L' Architecture d'Aujourd'hui*. All are available through direct subscription, from some bookshops or bookstalls, and may be found in college libraries.

Trade magazines (of which an astonishing number exist) are often very useful for their advertisements and other information.

Exchange and Mart and *Machinery Market* are useful for designers with workshop inclinations.

9   British Standards Institution: a source of continual reference for indispensable codes of practice, standards, and terminology, all of which are listed in the *Yearbook* or in sectional lists. There is also a full library for consultation at 2 Park Street, London W1. The library also keeps international standards. The work of the BSI is best described in the words of the *Yearbook:*

'The scope ... includes: glossaries of terms, definitions, quantities, units and symbols, methods of test; specifications for quality, safety, performance or dimensions; preferred sizes and types; codes of practice. The BSI is also concerned with certification and approval of products as complying with standards and with the international aspects of this subject.'

10   Design Centres: promoted by the Design Council, a government agency, and thus distinguished from the more specialized Building Centres, which are trade sponsored. The Council provides exhibitions of approved British design in manufacture, a design index for tracing well-designed retail products, a register of designers for industry, has a bookshop with a mail-order service, and publishes *Design* magazine. Not all designers see eye-to-eye with Council, but it should be remembered that a Design Centre is essentially a public institution.

11   Chartered Society of Designers (CSD): once the 'Society of Industrial Artists & Designers' (SIAD). A professional body similar in some respects to the RIBA (see 12). Issues scales of fees for designers (available to non-members), awards Membership or Fellowship as (in effect) a professional qualification, though this is not of course obligatory for the practice of design. Legal and information services to members. Has an educational role in awarding diploma membership to design students.

12   Royal Institute of British Architects (RIBA): the governing body for the regulation of architectural practice (e.g. it oversees education and professional qualifications, also forms of contract and fees). Has first-class library (open to the public for reference only) and a good bookshop, both at 66 Portland Place, London W1.

13   Exhibitions: in England the big trade exhibitions take place in the
National Exhibition Centre at Birmingham; in London – Earls Court,
Olympia, the Design Business Centre, are the main venues. Ask for their
programmes. Some exhibitions are particularly informative: the
Building Exhibition, for example, is very comprehensive and
provides a useful catalogue. 'Public' exhibitions (such as the Ideal
Home) are of less interest. Designers will often find unexpected
value in exhibitions quite off their normal beat, for instance the
Boat Show, the Mechanical Handling exhibition, Camping and
Caravaning . . . All these attract a hinterland of small exhibitors,
often with interesting products or services. It has to be said that the
duller exhibitions are occasions of business ritual (for swopping
notes, keeping an eye on each other, and drinking together) and have
little else to offer. Exhibition stands provide a source of design work
and are often interesting to see for that reason (the best seem to be
commissioned for the Building Exhibition).

Exhibitions of special interest to designers include those at the
Design Museum (see 35), the Institute of Contemporary Arts, the
Royal College of Art, the Design Centre, and occasionally elsewhere.
Art gallery information is outside the scope of these notes.

14   Consumers' guides: *Which?* is useful to designers who have to
specify products from the retail market; back issues are available in
libraries or may often be consulted in community centres or agencies
of the Consumers' Association.

15   Retail shops: a most obvious and useful source of information at
first hand. Also instructive in less obvious ways: furniture designers
and architects can learn much about how things go together, just by
looking. Warehouses, timber yards, scrap dealers, tool shops, and
surplus stores are all rich sources, often for the accidental find rather
than any more systematic purpose. Also useful for gathering samples
(of fabrics, timber offcuts, etc).

16   Trade representatives ('reps'): will often be only too happy to visit
designers and offer their (partial) advice and services, which will
include samples, literature, quotations, etc. This is their job and a
visit does not imply obligation.

17   Local authorities: a most useful and helpful source of information;
also of advice. Officers such as the district surveyor, fire officer,

officer of environmental health (ex 'sanitary inspector') will usually be pleased to discuss problems at an early stage, and *before* they cause difficulties.

18   Museums: useful for their exhibitions, for research, and for their specialist libraries. In London, the Victoria and Albert, the Science Museum, and the British Museum are, of course, indispensable, but there are also smaller specialist museums (e.g. the Geffrye), and many based on local history (e.g. the Great Western Railway Museum at Swindon, or, for industrial archaeology, the Ironbridge Gorge Museum at Telford. The Design Museum (see 35) is a fresh addition.

19   Colleagues, mates, friends, even enemies – may all be helpful sources of information, involuntary or otherwise. Sometimes a neglected source.

20   Design and Industries Association (DIA): a non-professional body which campaigns for improved standards of design, and organizes lectures and exhibitions; active regionally.

21   Open University (OU): provides a growing number of books and course documents and related radio and TV programmes; available to non-students and fully catalogued by OU Educational Enterprises.

22   The Advisory Centre for Education (ACE) and the Council for Educational Technology (CET): worth knowing for their lists and services; the latter may advise, for instance, with regard to Prestel, the Post Office view-data system, and other electronic aids to learning.

23   Council for Small Industries in Rural Areas (CoSIRA): formerly the Rural Industries Bureau with a distinguished record of help to small producers and workshops, and to service trades. Government supported but independent. Can offer advice over a wide range of practical matters from book-keeping to machine maintenance. Arranges instruction by area visits or centrally and there are loan schemes. Liaison officers in most English counties.

24   Department of the Environment (DoE): umbrella institution responsible for the control and protection of environment, concerned with all forms of pollution, road research, etc; many devolved organizations (see 25). Large specialized library available to negotiated use by students.

25 Building Research Establishment (BRE): combines what was formerly the Building Research Station, the Fire Research Station, and the Forest Products Research Laboratory (now known as the Princes Risborough Laboratory of the BRE). Since each maintains its own independent advisory, research, and publications service, and their own addresses, this amalgamation has caused some confusion. Annual information directory.

26 Princes Risborough Laboratory of the Building Research Establishment (formerly the Forest Products Research Laboratory): publishes the standard descriptive references for timber and conducts research at that level; publications list and services advice available. Not to be confused with TRADA (see 27).

27 Timber Research and Development Association (TRADA): a trade sponsored association concerned mainly with the uses of timber, though there is some overlap with Princes Risborough (see 26). Local offices and advisory services with exhibits at Building Centres. Publications and services lists.

28 Furniture Industry Research Association (FIRA): directs research into the technical aspects of the furniture industry and publishes books and research papers; lists available; occasional courses.

29 Small Firms Service: run by the Employment Department; provides advice, initially free of charge, on any business problem, through local Small Firms Centres. Publications include *Starting your own business* and *Accounting for a small firm*. Dial 100 and phone 'Freephone Enterprise'.

30 Crafts: the Crafts Council (formerly the Crafts Advisory Committee) is a government-funded body that supports and promotes the crafts in England and Wales (there is a Scottish equivalent). Among its services: grants and loans, a register of craftsmen, slide library, exhibitions (touring and at their Waterloo Place gallery), publications – including *Crafts* magazine and the useful booklet, *Running a workshop*. Contemporary Applied Arts (an independent organization) presents regular exhibitions in its Covent Garden Gallery. The Society of Designer Craftsmen is a national umbrella body.

31   Alternatives: among the groups working on applications of alternative technology are the Centre for Alternative Technology (CAT), the Intermediate Technology Development Group (ITDG), and at the Faculty of Technology within the Open University. A promising initiative has been the Centre for Alternative Industrial and Technological Systems (CAITS), working with an approach argued for by Mike Cooley (see his *Architect or bee?*). Friends of the Earth (FoE) are best known among the campaigning bodies.

32   Listings magazines: calendars of events published in the metropolitan centres (e.g. *Time Out* in London) can provide useful information hard to find elsewhere.

33   Trade and professional institutes and associations: too numerous and specialized to list here (other than the more obvious ones given above), but a most valuable source of technical advice, publications, and other services, which may include conferences and short refresher courses to keep technical information up-to-date. Publications will include books, research papers, information digests, and notes to aid specification. Students are strongly advised to investigate the whole range of availability and to obtain particulars (usually gratis) of all areas that concern them. Publications and services may also, in some cases, be without charge, when an association is a promotional body and trade-sponsored. Addresses and short descriptions are given in the *Building Centre and CIRIA guide to sources of information* (in the construction industry), which is much wider ranging than its title suggests, and/or in the directories of associations and professional bodies, normally available in a library.

34   Pocket books, diaries, etc, often have quick-reference tables and other handy information. An example of a good trade pocket book is Charles Hayward's *Woodworker's pocket book*. The Lefax (leaves plus facts) is an information system of considerable scope and ingenuity. All in all, an interesting example of a design solution that develops a system, and a process, out of finite objects like diaries and cash wallets – with which, of course, it offers an instructive contrast. It is also interesting that the conceptually superseded lesser functions that the Lefax subsumes, are often preferred in their original form, the Lefax being too large, complicated, and perhaps offering too many worrying choices (see remarks about flexibility in building design in part 6). A useful test case for experiment and discussion.

Since 1980, when the paragraph above was written, Lefax and Filofax have become separate concerns, and their products (and those of imitators) are popular and widely available. With the mushrooming of 'designer' this-and-that, the Filofax, particularly, has become an essential stage-property for any successfully competitive life-style. Not quite what the doctor ordered – as a clarifying agent in our confusion of urbanized signal inputs – and still far from 'optimal' as a system, but still good for graphic projects (and for critical analysis by socially-aware design students).

35   The Design Museum: a 'museum of everyday things', established and run by the Conran Foundation as an educational venture, with sponsorship from industry and some public funding. Activities include exhibitions and a review of products; there is also a library, shop, restaurant and café.

19   Using libraries

The use of libraries is a special case of 'finding out for yourself' (part 18). Library systems are complex and sometimes baffling to the uninitiated, but their mysteries yield to a little knowledge of the principles on which their services are organized. Obviously this is no substitute for expert advice from a librarian: when in difficulty, always ask. The first thing is to know the overall pattern of availability in a search for information. Then a simple outline of the normal classifications should be kept in view, if not in mind. It gives confidence to know that in principle, libraries do interconnect and therefore nearly all books are ultimately obtainable (given time). There is some variation in the way libraries card-index or otherwise identify their stock; traditional methods may be augmented or replaced by microfiche and other systems for the storing and retrieval of information. Although classification systems are nationally and internationally recognized, the 'order' by which a library becomes fully intelligible, and workable, is the responsibility of the librarian. The larger libraries issue leaflets to explain in detail how they are organized. Normally however it is unwise to rely on a simple familiarity with shelf-order. A librarian may be able to suggest alternatives, and should certainly be consulted for periodical indexes, cuttings files, microform, and other sources.

The notes that follow are in five parts:

1   Types of library

2   Classification systems

3   Finding books

4   Finding information

5   Special services

1   Types of library

Public libraries: a widespread and familiar service with effective
support from the interloan system, which operates through the British
Library to specialized sources both in this country and abroad, in
addition to interlinked regional resources. Loan and reference; open
to all.

Educational libraries: ranging in scope from the modest libraries for
schools to the major collections of the universities. College libraries
will normally specialize on a faculty basis, as for art and design, and
will use for that reason fine subdivisions of their subject
classification. Slide, film, cassette, video-cassette, and copying
services will be available and there will usually be specialized
magazine and periodical collections. Open to students.

Special libraries: numerous and often highly specialized. Includes
the libraries of museums, government and trade institutes, industry,
and the private libraries. Some are accessible by members'
subscription, some for research purposes only, while others (like the
RIBA Library) are open to the public for reference only. Listed in
the *Libraries, museums, and art galleries year book,* and in the *Aslib
directory.* Before making special arrangements to use such libraries
it is worth checking if they connect with an interloan scheme.

Source libraries: national storage systems now embraced by the
British Library (Reference and Lending Divisions) which receives
copies of every book published in Britain. There are national library
catalogues, sometimes with subject indexes (e.g. American
publications are covered by the *National union catalog*). When
books are not available on loan, enquire whether there is a
photocopying service. Libraries of this calibre are of most concern
to librarians and to serious research scholars.

2   Classification systems

Dewey decimal system: as used by public and college libraries.
Divides knowledge into 10 classes, then further subdivides by 10,
and so on, working from the general to the particular. After the
third digit a decimal point is used for clarification.

Main divisions		Arts subdivisions	
000	general	700	general
100	philosophy / psychology	710	planning
200	religion	720	architecture
300	social sciences	730	sculpture / ceramics
400	language	740	design / drawing
500	science	750	painting
600	technology	760	printmaking
700	arts	770	photography
800	literature	780	music
900	history / geography	790	recreation (e.g. sport, cinema)

A common example is as follows:

800	Literature
820	(Literature) English
822	(Literature) (English) Drama
822.3	(Literature) (English) (Drama) Elizabethan
822.33	(Literature) (English) (Drama) (Elizabethan) Shakespeare

from which it will be seen that the placing of the decimal point is an arbitrary (non-logical) convention. Designers may find themselves most often in the 700 ranges; an example here would be 'English architecture':

700	Art
720	(Art) Architecture
720.942	(Art) (Architecture) English

from which a certain clumsiness becomes apparent. For this and other reasons, scientific and technical libraries, including those used by designers, may employ a development of Dewey known as UDC (Universal Decimal Classification). UDC uses distinctive notations for component features, which can be combined to form more concise descriptions of a complex subject. Designers should also be aware of the CI/SfB system of classification; the initials stand for 'Construction Index' and 'Samarbetskommittén för Byggnadsfrågor' (the name of the classification system of the Swedish construction industry). This provides a system of classification and notation for all information in the subject-field of construction, building, and architecture, and much published technical information is pre-classified according to CI/SfB.

In practice, classification systems are easier to use than to describe. Dewey, UDC, and CI/SfB are all served by published schedules and tables, which incorporate alphabetical subject indexes. Look here for the particular heading that you want, not the general heading (or you may never get there).

3   Finding books

If you are looking for a 'good book' on a subject, this may be an individual matter, but in specialized studies (such as art and design) there are recognizable 'standard works' and others known to be of special interest, perhaps for their authenticity, originality of outlook, special knowledge, or other qualities. There are four ways of finding such books. The first is to ask someone whose judgement and experience you have reason to trust. The second is to work outwards from a bibliography as is given in part 8 ('Reading for design'), or from bibliographies given elsewhere (as in Lewis Mumford's trilogy mentioned in that part and as in many similar books). The third is to follow critical recommendations in the journals. Somewhat unreliable, because the instant-culture pressures of critical journalism tend to give an over-generous reception to new books, and there is rarely enough space to estimate their contribution against established sources. Anyone who looks through book reviews that are a few years old will see that book-reviewing is a culture of its own, and often a substitute for book reading rather than a guide to it. The fourth way is to go to a library or bookshop, find the right classification, and browse through the shelves. Although this method has the benefit of immediacy – the book can actually be seen and handled – it is the least satisfactory. The best books may be out on loan; or, in a bookshop, may be either sold out or temporarily out of print. It is also a sad fact of contemporary publishing that the most glamorous-looking productions often turn out to be the most derivative or insubstantial. Obvious as these remarks may be, it is surprising how many students rely on this hit-and-miss procedure to govern their reading, and by such means, often miss some of the best books on their subject. If it is a matter for systematic research, or for the *forming* of a select bibliography (which may amount to the same thing), or if it is necessary to *trace* a book, the following sources should be consulted:

British book in print: published in two volumes annually, lists against author and title all British books in print at the time of publication. Available on (and updated by) microfiche.

British national bibliography: lists all books published in Britain; cumulative. Published weekly; indexes cumulate in each monthly part; whole volume cumulates annually. Based on the British Library. Authors, titles, and subjects.

Cumulative book index: an American publication listing books published in the English language irrespective of their country of origin. Published monthly and as the title suggests, cumulative. Alphabetical listing for subject, author, and title.

Catalogues of national libraries: extensive coverage for the country concerned; of scholarly interest but less likely to concern designers.

Subject bibliographies: may appear from time to time from various sources (including books as mentioned above); consult a librarian and see:

World bibliography of bibliographies (Besterman): alphabetical subject heading; an index to a vast number of bibliographies on all subjects and in many languages. The ultimate reference in book form: to 1966; supplemented by A.F.Toomey for 1964-74.

4   Finding information

If the subject is unfamiliar, it is worth beginning with a good (i.e. authoritative) encyclopaedia, or encyclopaedic-dictionaries as they become in more specialized fields. There will usually be a short standard bibliography or a booklist without annotation. The recommendations may be too 'safe' to meet your requirements; on the other hand, any reference given special mention should at least be consulted. If the subject is familiar, a library can extend your knowledge either through books (see previous advice), through magazines, which are stored cumulatively in specialist libraries, through indexed guides to information including illustrative material, and through access to slides, microform, etc. (see 5: Special services). If the query is merely factual, there are many directories, almanacs, dictionaries, and biographical dictionaries. If you are uncertain, it is worth repeating: ask the librarian.

Encyclopaedia Brittanica: American in origin, despite its name. Too well known (and justly so) to require comment. Note the date of the edition you consult; out-of-date information can be a difficulty in using encyclopaedias, and one reason why they should only be turned to for a very basic account of a subject.

Chambers encyclopaedia: a British alternative to the above.

Subject encyclopaedias: there are many, and new ones appear fairly frequently. Examples: encyclopaedias of modern architecture, world art, typefaces, etc. Some specialist dictionaries are in effect encyclopaedic (e.g. *Dictionary of art and artists*). It must be said that the subject encyclopaedias are uneven in quality.

Biographical dictionaries: usually national in outlook, e.g. the well-known *Dictionary of national biography.* There are more concise alternatives. For current reference there are *Who's who* (general), *Who's who in education,* and so on, though students should know that not everyone agrees to be listed.

Year books: often useful for quick reference. Examples are the *Libraries, museums, and art galleries year book* (already mentioned), and the *Writers' and artists' year book* and *Educational year book,* each of which has facts, statistics, addresses, etc, of interest to designers. The British Standards Institution *Yearbook* is, in effect, their catalogue. *Spon's architects' and builders' price guide* and similar tables must obviously be as up-to-date as possible.

Directories: telephone directories are indispensable (A/Z and classified Yellow Pages). There are many specialized volumes and it is worth finding out what they have to offer. Relevant examples: *Aslib directory* (listing special libraries), *Kelly's manufacturers and merchants directory, Kompass: register of British industry and commerce,* Millard's *Trade associations and professional bodies of the United Kingdom, Directory to the furnishing trade.* When in doubt, ask a librarian.

Almanacs: *Whitaker's* gives many facts and figures and is worth scanning to estimate its possible usefulness (otherwise it will remain unopened). Specialized almanacs include the redoubtable *Reed's* and *Brown's* for nautical reference.

Indexes to articles in periodicals: the Library Association issues the *British humanities index* and the *British technology index* (British periodical literature only). There are also indexes that cover the American journals and some foreign-language sources. The *Architectural periodicals index* (published by the RIBA) should be given special mention. The *Art index* is useful. Consult a librarian for help in this area.

Periodicals: see above, see also *Willing's press guide* (British) and *Ulrich's international periodicals directory*. The specialist libraries will keep considerable reserves of ephemera which does, of course, eventually acquire a patina of history.

Dictionaries: *Chambers* is a good single-volume dictionary; the *Oxford English dictionary* is of course a great work, the two-volume *Shorter Oxford* being the next best thing. There are many specialist dictionaries (e.g. of slang, of quotations), those of foreign languages, and those which are abbreviated encyclopaedias (qv). The English-language dictionaries vary in their bias (some being more technical or American than others) and in their method – the OED even in its shortest versions is more contextual than others. Librarians should be consulted if such distinctions are felt to be relevant.

Language aids: for the general reader and writer, and for designers in those respect, Fowler's *Dictionary of modern English usage* is the standard work, updated by Partridge's *Usage and abusage*. There are many more specialized reference books. Dictionaries of synonyms and antonyms (such as Roget's *Thesaurus*) may be found helpful. There are 'how to' manuals (e.g. to read, write, study) which should be approached with caution.

5   Special services

It is impossible to generalize here, but students should be aware that libraries do not only store books and periodicals, they may offer microfiche and microfilm facilities, video-tapes and possibly (with due notice) a recording service from radio or TV, copying facilities, slide and film collections, Open University information and material, and some libraries may organize their own lecture programmes, some may show films, and some may lend or play recorded music. These matters should be enquired into in all the specialized libraries. Libraries may offer computer searching of abstracts and indexes. Such services are apt to be expensive but may interest researchers and post-graduate workers.

The list comprises most titles mentioned in this book. As should be clear from the introductory remarks in 'Reading for design' (part 8), this does not offer a full and balanced recommendation for design reading, and it is also far too long for that purpose. Students will find their own priorities, but the list should be helpful for reference. Details of the first British edition are generally given (place of publication London, unless otherwise indicated), with additional reference to recent paperback editions.

ABERCROMBIE, M.L.J, *The anatomy of judgement*, Hutchinson, 1960. (Harmondsworth: Penguin Books, 1969)

ALEXANDER, Christopher, *Notes on the synthesis of form*, Cambridge Mass: Harvard University Press, 1964

— *The timeless way of building*, New York: Oxford University Press, 1979

ANARCHY (magazine), ed. Colin Ward, 1961-70

ARCHER, L. Bruce, *Systematic method for designers*, Council of Industrial Design, 1965

ASHBEE, C.R, *Craftsmanship in competitive industry*, Campden: Essex House Press, 1908

— *Should we stop teaching art?* Batsford, 1911

ASSOCIATION FOR PLANNING AND REGIONAL RECONSTRUCTION, *Survey before plan* (series), Lund Humphries, [1945]

AYLWARD, Bernard, *Design education in schools*, Evans, 1973

BALLANTINE, Richard, *Richard's bicycle book*, Pan Books, 1975

BAUHAUS BOOKS, nos. 1-14, Munich: Albert Langen, 1925-30. (Facsimiles and new editions, Mainz/Berlin: Florian Kupferberg, 1965-; some English editions, Lund Humphries)

BAYER, Herbert, and Walter Gropius, Ise Gropius, (ed.) *Bauhaus 1919-1928*, New York: Museum of Modern Art, 1938. (Secker & Warburg, 1976)

BEAZLEY, Elisabeth, *Design and detail of the space between buildings*, Architectural Press, 1960

BENEDICT, Ruth, *Patterns of culture*, Routledge & Kegan Paul, 1935

BENTON, Tim, and Charlotte Benton, with Dennis Sharp, (ed.) *Form and function*, Crosby Lockwood Staples, 1975

BERNERI, M.L, *Journey through Utopia*, Routledge & Kegan Paul, 1950

BOARDMAN, Philip, *The worlds of Patrick Geddes*, Routledge & Kegan Paul, 1978

BUBER, Martin, *Between man and man*, Routledge & Kegan Paul, 1947. (Fontana, 1961)

— *The way of man*, Routledge & Kegan Paul, 1950. (Vincent Stuart, 1963)

— *Paths in Utopia*, Boston: Beacon Press, 1958

CANTACUZINO, Sherban, *Wells Coates*, Gordon Fraser, 1978

CAPELL, Richard, *Schubert's songs*, 3rd edn, Pan Books, 1973. (Duckworth, 1974)

COOLEY, Mike, *Architect or bee? the human price of technology*, 2nd edn, Hogarth Press, 1987

COX, C.B, and A.E.Dyson, Rhodes Boyson, (ed.) *Black papers on education*, nos. 1-5, 1969-77

CRESWELL, H.B, *The Honeywood file*, Architectural Press, 1929

CROSS, Nigel, David Elliott and Robin Roy, (ed.) *Man-made futures*, Hutchinson, 1974

CULLEN, Gordon, *The concise townscape*, Architectural Press, 1971

CURTIS, William J.R, *Modern architecture since 1900*, 2nd edn, Oxford: Phaidon, 1987

DAVIE, Donald, *Articulate energy: an enquiry into the syntax of English poetry*, Routledge & Kegan Paul, 1955

DOMUS BOOKS (Quaderni di Domus), Milan: Editoriale Domus,

DUBERMAN, Martin, *Black Mountain: an exploration in community*, Wildwood House, 1974

EAGLETON, Terry, *Literary theory*, Oxford: Blackwell, 1983

ECOLOGIST, THE, (editors of) *A blueprint for survival*, Harmondsworth: Penguin Books, 1972

FORM (magazine), Cambridge, 1966-9

FRAMPTON, Kenneth, *Modern architecture: a critical history*, Thames & Hudson, 1980

FINCH, Janet and Michael Rustin, (ed.) *A degree of choice: higher education: a right to learn*, Harmondsworth: Penguin Books, 1986

FROMM, Erich, *Fear of freedom*, Routledge & Kegan Paul, 1942

— *The sane society*, Routledge & Kegan Paul, 1956

FROSHAUG, Anthony, *Typographic norms,* Birmingham: Kynoch Press (with Design & Art Directors Association, London), 1964

GABO, Naum, and J.L.Martin, Ben Nicholson, (ed.) *Circle,* Faber & Faber, 1937. (Facsimile edition, Faber & Faber, 1971)

GARLAND, Ken, *Illustrated graphics glossary,* Barrie & Jenkins, 1980

GHYKA, Matila, *A practical handbook of geometrical composition and design,* Tiranti, 1956

GIEDION, Sigfried, *Space, time and architecture* (1941) 5th edn, Cambridge Mass: Harvard University Press, 1967

— *Mechanization takes command,* New York: Oxford University Press, 1948

— *The eternal present* trilogy (New York: Oxford University Press): *The beginnings of art* (1962); *The beginnings of architecture* (1964); *Architecture and the phenomena of transition* (Cambridge Mass: Harvard University Press, 1971)

GOFFMAN, Erving, *Encounters,* Indianapolis: Bobbs Merrill, 1961

— *Interaction ritual,* Chicago: Aldine, 1967 (and other books)

GOODMAN, Paul, *Compulsory miseducation,* Harmondsworth: Penguin Books, 1971

GOODMAN, Percival, and Paul Goodman, *Communitas,* Chicago: Chicago University Press, 1947. (New York: Vintage Books, 1973)

GOSLETT, Dorothy, *The professional practice of design,* 2nd edn, Batsford, 1978

GREEN, Ronald, *The architect's guide to running a job,* 3rd edn, Architectural Press, 1972

GROPIUS, Walter, *The new architecture and the Bauhaus,* Faber & Faber, 1935

GUINNESS, Oswald, *The dust of death,* Leicester: Inter-Varsity, 1973

HAYWARD, Charles, *Woodworker's pocket book,* Evans, 1971

HORNSEY COLLEGE OF ART (students and staff), *The Hornsey affair,* Harmondsworth: Penguin Books, 1969

ILLICH, Ivan D, *Deschooling society,* Harmondsworth: Penguin Books, 1971

JENCKS, Charles, *The language of post-modern architecture,* 2nd edn, Academy Editions, 1978

JENCKS, Charles, and George Baird, (ed.) *Meaning in architecture*, Barrie & Rockliff, 1969

JONES, J. Christopher, *Design methods*, Chichester: Wiley, 1970

KING, Anthony D, (ed.) *Buildings and society*, Routledge & Kegan Paul, 1980

KITCHEN, Paddy, *A most unsettling person: an introduction to the ideas and life of Patrick Geddes*, Gollancz, 1975

KOESTLER, Arthur, *The act of creation*, Hutchinson, 1964. (Pan Books, 1975)

KOFFKA, Kurt, *Principles of Gestalt psychology*, Routledge & Kegan Paul, 1935. (New York: Harcourt Brace, 1967)

KÖHLER, Wolfgang, *Gestalt psychology*, Bell, 1930. (New York: Mentor Books, 1966)

— *The place of value in a world of facts*, Routledge & Kegan Paul, 1939. (New York: Liveright, 1976)

KRON, Joan, and Suzanne Slesin, *High-tech*, Allen Lane, 1979

KROPOTKIN, Peter, *Fields, factories and workshops*, Hutchinson, 1899. (Edited and with new material, by Colin Ward: *Fields, factories and workshops tomorrow*, George Allen & Unwin, 1974)

— *Mutual aid: a factor of evolution*, Heinemann, 1902. (Allen Lane, 1972, edited by Paul Avrich)

LE CORBUSIER, *Towards a new architecture*, John Rodker, 1927. (Facsimile edition, Architectural Press, 1970)

— *Œuvre complète* (ed. W.Boesiger), Zürich: Girsberger, 1929-

LETHABY, W.R, *Form in civilization* (1922) 2nd edn, Oxford University Press, 1957

— aphorisms, see Roberts (below)

LISSITSKY-KÜPPERS, Sophie, *El Lissitsky: life, letters, texts*, Thames & Hudson, 1968

LOGIE, Gordon, *Furniture from machines*, George Allen & Unwin, 1948

MACEWAN, Malcolm, *Crisis in architecture*, RIBA Publications, 1974

MARTIN, Bruce, *Standards and building*, RIBA Publications 1971

— *Joints in buildings*, George Godwin, 1977

MARTIN, J.L, and S.Speight, *The flat book*, Heinemann, 1939

MATRIX, *Making space: women and the man-made environment*, Pluto Press, 1984

MEDAWAR, Peter, *The art of the soluble*, Methuen, 1967

— *Induction and intuition in scientific thought*, Methuen, 1969

MITCHELL, Percy, *A boatbuilder's story,* St Austell: Kingston Publications, 1968

MOHOLY-NAGY, László, *The new vision,* 3rd edn, New York: Wittenborn, 1946

MUMFORD Lewis, *Technics and civilization,* Routledge & Kegan Paul, 1934. (New York: Harcourt Brace, 1975)

— *The culture of cities,* Secker & Warburg, 1938. (New York: Harcourt Brace, 1970)

— *The condition of man,* Secker & Warburg, 1944. (New York: Harcourt Brace, 1973)

NAIRN, Tom, and Angelo Quattrocchi, *The beginning of the end,* Panther Books, 1968

NEEDHAM, Joseph, *Science and civilisation in China,* Cambridge: Cambridge University Press, 1954-

NEILL, A.S, *Summerhill: a radical approach to education,* Gollancz, 1962. (Harmondsworth: Penguin Books, 1970)

OPEN UNIVERSITY, *History of architecture and design 1890-1939* (A305: course units), Milton Keynes, Open University Press, 1975

— *Man-made futures: design and technology* (T262: course units), Milton Keynes: Open University Press, 1975

PAPANEK, Victor, *Design for the real world,* Thames & Hudson, 1972.

PATEMAN, Trevor, (ed.) *Counter course: a handbook for course criticism,* Harmondsworth: Penguin Books, 1972

PETERS, R.S, *Ethics and education,* George Allen & Unwin, 1966 (and other books)

PEVSNER, Nikolaus, *Pioneers of modern design,* 3rd edn, Harmondsworth: Penguin Books, 1960

PIRSIG, Robert M, *Zen and the art of motor cycle maintenance,* Bodley Head, 1974. (Corgi Books, 1976)

POUND, Ezra, *ABC of reading,* Faber & Faber, 1951

PYE, David, *The nature and art of workmanship,* Cambridge: Cambridge University Press, 1968

RASCH, Heinz, and Bodo Rasch, *Der Stuhl,* Stuttgart: Akademischer Verlag Fritz Wedekind, ?1929

READ, Herbert, *Art and industry* (1934) 4th edn, Faber & Faber, 1956

— *Anarchy and order,* Faber & Faber, 1945

— *Education through art,* revised edn, Faber & Faber, 1958

RÉE, Harry, *Educator extraordinary: the life and achievement of Henry Morris 1889-1961*, Longman, 1973

REICH, Wilhelm, *The function of the orgasm*, Panther Books, 1968

REIMER, Everett, *School is dead: an essay on alternatives*, Harmondsworth: Penguin Books, 1971

Report from the Select Committee on Education and Science: session 1968-69: student relations (7 vols.), HMSO, 1969

RICHARDS, I.A, *Practical criticism*, Routledge & Kegan Paul, 1929

ROBERTS, A.R.N, *William Richard Lethaby 1857–1931*, LCC Central School of Arts & Crafts, 1957

ROSZAK, Theodore, *The making of a counter-culture*, Faber & Faber, 1970

ROTH, Alfred, (ed.) *The new architecture*, Zürich: Les Éditions d'Architecture, 1940

SALAMAN, R.A, *Dictionary of tools, used in the woodworking and allied trades, c.1700-1970*, George Allen & Unwin, 1975

SAMSON, Frederic, *Dotes and antidotes*, Royal College of Art, 1979

SCHUMACHER, E.F, *Small is beautiful*, Blond & Briggs, 1973. (Sphere Books, 1974)

— *A guide for the perplexed*, Blond & Briggs, 1977. (Sphere Books, 1978)

—*Good work*, Jonathan Cape, 1979. (Sphere Books, 1980)

SHARP, Dennis, (ed.) *Planning and architecture: essays presented to Arthur Korn by the Architectural Association*, Barrie & Rockliff, 1967

— (ed.) *The rationalists*, Architectural Press, 1978

SKELTON, Robin, (ed.) *Herbert Read: a memorial symposium*, Methuen, 1970

SMITHSON, Peter, *Bath: walks within walls*, Bath: Adams & Dart, 1971

SOMMER, Robert, *Tight spaces: hard architecture and how to humanize it*, Englewood Cliffs: Prentice-Hall, 1974

SPENCER, Herbert, *Pioneers of modern typography*, Lund Humphries, 1969

STEINER, George, *Language and silence*, Faber & Faber, 1967. (Harmondsworth: Penguin Books, 1969)

STURT, George, *The wheelwright's shop*, Cambridge: Cambridge University Press, 1923

THOMPSON, E.P, *William Morris: romantic to revolutionary* (1955) revised edn, Merlin Press, 1977

TOFFLER, Alvin, *Future shock*, Bodley Head, 1970. (Pan Books, 1973).

TSCHICHOLD, Jan, *Typographische Gestaltung*, Basel: Benno Schwabe, 1935. (English edition: *Asymmetric typography*, Faber & Faber, 1967)

TURNER, John F.C, *Housing by people*, Marion Boyars, 1976

ULM (magazine), Ulm: Hochschule für Gestaltung, 1958-68

WACHSMANN, Konrad, *The turning point of building*, New York: Van Nostrand Reinhold, 1961

WARD, Colin, *The child in the city*, Architectural Press, 1978

— (ed.) *Vandalism*, Architectural Press, 1973

WATKIN, David, *Morality and architecture*, Oxford: Oxford University Press, 1977

WHITEHEAD. A.N, *The aims of education*, Williams & Norgate, 1929. (Benn, 1949)

WHOLE EARTH CATALOG, *The last Whole Earth Catalog*, Harmondsworth: Penguin Books, 1971

— *Whole Earth epilog*, Harmondsworth: Penguin Books, 1974

WHYTE, L.L, (ed.) *Aspects of form* (1951) 2nd edn, Lund Humphries, 1968

WINGLER, Hans M, *The Bauhaus*, Cambridge Mass: MIT Press, 1969

WOODCOCK, George, *Herbert Read: the stream and the source*, Faber & Faber, 1972

WOODEN BOAT (magazine), PO Box 4943, Manchester, NH 03102, USA

ZEVI, Bruno, *The modern language of architecture*, Washington: University of Washington Press, 1978

addenda (see part 8)

CRAWFORD, Alan, *C. R. Ashbee*, New Haven: Yale University Press, 1985

RUBENS, Godfrey, *William Richard Lethaby*, Architectural Press, 1986

21 Useful addresses (GB)

ACE Advisory Centre for Education
18 Victoria Park Square, London E2 9PB
01-980 4596

AA Architectural Association School of Architecture
34-36 Bedford Square, London WC1B 3ES
01-636 0974

Art Workers Guild
6 Queen Square, London WC1 3AR
01-837 3474

Arts Council of Great Britain
105 Piccadilly, London WC1V 0AU
01-629 9495

Aslib Association for Information Management
26-27 Boswell Street, London WC1N 3JZ
01-430 2671

Blackwell's (B.H.Blackwell Ltd)
Broad Street, Oxford OX1 3BQ
Oxford (0865) 49111

British Library Reference Division
Great Russell Street, London WC1B 3DG
01-636 1544

British Museum
Great Russell Street, London WC1B 3DG
01-636 1555

BSI British Standards Institution
2 Park Street, London W1A 2BS
01-629 9000

Building Centre
26 Store Street, London WC1E 7BT
01-637 1022/3151 (bookshop)

BRE   Building Research Establishment
Building Research Station, Garston, Watford WD2 7JR
Garston (0923) 674040

BRE: Fire Research Station
Melrose Avenue, Borehamwood, Herts WD6 2BL
01-953 6177

BRE: Princes Risborough Laboratory
Aylesbury, Bucks HP17 9PX
Princes Risborough (084 44) 3101

Business Design Centre
52 Upper Street, London N1 0QH
01-359 3535

B/TEC   Business & Technician Education Council
Central House, Upper Woburn Place, London WC1H 0HH
01-388 3288

Central Bureau for Educational Visits & Exchanges
Seymour Mews House, Seymour Mews, London W1H 9PE
01-486 5101
(also Scottish and Irish offices)

CAITS   Centre for Alternative Industrial and Technological Systems
Polytechnic of North London, Holloway Road, London N7 8DB
01-607 7079

Centre for Alternative Technology
Llwyngwern Quarry, Machynlleth, Powys SY20 9AZ
Machynlleth (0654) 2400

CSD   Chartered Society of Designers
29 Bedford Square, London WC1B 3EG
01-631 1510

City and Guilds of London Institute
76 Portland Place, London W1N 4AA
01-580 3050

Collet's International Bookshop
129 Charing Cross Road, London WC2H 0EQ
01-734 0782

Compendium Bookshop
234 Camden High Street, London NW1 5XQ
01-485 8944

CIRIA   Construction Industry Research & Information Centre
6 Storey's Gate, London SW1P 3AU
01-222 8891

Consumers' Association
14 Buckingham Street, London WC2N 6DS
01-839 1222

Contemporary Applied Arts
43 Earlham Street, London WC2H 9LD
01-836 6993

CAFD   Council for Academic Freedom & Democracy
21 Tabard Street, London SE1 4LA
01-403 3888

Council for Educational Technology
3 Devonshire Street, London W1N 2BA
01-580 7553

CNAA   Council for National Academic Awards
344-354 Gray's Inn Road, London WC1X 8BP
01-278 4411

CoSIRA   Council for Small Industries in Rural Areas
141 Castle Street, Salisbury, Wilts SP1 3TB
Salisbury (0722) 336255

Crafts Council
12 Waterloo Place, London SW1Y 4AU
01-930 4811

DES Department of Education and Science
Elizabeth House, York Road, London SE1 7PH
01-934 9000

DoE Department of the Environment
2 Marsham Street, London SW1P 3EB
01-212 3434

Design Centre/
Design Centre Bookshop/
Design Council
28 Haymarket, London SW1Y 4SU
01-839 8000

DHS Design History Society
Journal: Oxford University Press, Walton Street,
Oxford OX2 6DP

DIA Design and Industries Association
Secretary, 17 Lawn Crescent, Kew Gardens, Surrey TW9 3NR
01-940 4925

Design Museum
Butler's Wharf, Shad Thames, London SE1 2YD
01-407 6265

DRS Design Research Society
Secretary, c/o RCA, Kensington Gore, London SW7 2EU

Dillon's Bookstore
1 Malet Street, London WC1E 7JB
01-636 1577
Dillon's Arts Bookshop
8 Long Acre, London WC2E 9LG
01-836 1359

Earls Court Exhibition Centre
Warwick Road, London SW5 9TA
01-385 1200

Freedom Press Bookshop
84b Whitechapel High Street, London E1 7QX
01-247 9249

FoE   Friends of the Earth
26 Underwood Street, London N1 7JQ
01-490 1555

FIRA   Furniture Industry Research Association
Maxwell Road, Stevenage SG1 3EW
Stevenage (0438) 313433

Geffrye Museum
Kingsland Road, London E2 8EA
01-739 8368

Heffer's (W.Heffer & Sons Ltd)
20 Trinity Street, Cambridge CB2 3NG
Cambridge (0223) 358351

Housmans Bookshop
5 Caledonian Road, London N1 9DX
01-837 4473

ICA   Institute of Contemporary Arts
12 Carlton House Terrace, London SW1Y 5AH
01-930 0493

ITDG   Intermediate Technology Development Group
Myson House, Railway Terrace, Rugby, Warwicks CV21 3HT
Rugby (0788) 60631

ICOGRADA   International Council of Graphic Design Associations
PO Box 398, London W11 4UG
01-603 8494

ICSID   International Council of Societies of Industrial Design
Kluvikatu 1d, 00100 Helsinki, Finland
626 661

IFI   International Federation of Interior Designers
PO Box 19610, 1000 GP Amsterdam, Netherlands

Ironbridge Gorge Museum
Ironbridge, Telford, Salop TF8 7AW
Ironbridge (095 245) 3522

Library Association
7 Ridgemount Street, London WC1E 7AE
01-636 7543

NADE   National Association for Design Education
Secretary, Kirbyhill, Plawsworth, Chester le Street,
County Durham DH2 3LD

NATFHE   National Association of Teachers in Further and Higher Education
15 Britannia Street, London WC1X 9JP
01-837 3636

NCCL   National Council for Civil Liberties
21 Tabard Street, London SE1 4LA
01-403 3888

NEC   National Exhibition Centre
Birmingham B40 1NT
021-780 4141

National Institute of Adult Education
19b De Montfort Street, Leicester LE1 7GE
Leicester (0533) 551451

NSEAD   National Society for Education in Art & Design
Secretary, 7a High Street, Corsham, Wilts SN13 0ES

NUS   National Union of Students
461 Holloway Road, London N7 6LJ
01-272 8900

Olympia
Hammersmith Road, London W14 8UX
01-603 2141

Open University
Walton Hall, Milton Keynes MK7 6AA
Milton Keynes (0908) 74066

Open University Educational Enterprises Ltd
12 Cofferidge Close, Stony Stratford, Milton Keynes MK11 1BY
Milton Keynes (0908) 566744

PIRA [The Research Association for the Paper and Board, Printing and
        Packaging Industries]
      Randalls Road, Leatherhead, Surrey KT22 7RU
      Leatherhead (0372) 76161

RA   Royal Academy of Arts
      Burlington House, Piccadilly, London W1V 0DS
      01-734 9052

RCA   Royal College of Art
      Kensington Gore, London SW7 2EU
      01-584 5020

RIBA Royal Institute of British Architects
      66 Portland Place, London W1N 4AD
      01-580 5533

RSA   Royal Society of Arts
      8 John Adam Street, London WC2N 6EZ
      01-930 5115

      Schools Curriculum Development Committee
      Newcombe House, 45 Notting Hill Gate, London W11 3JB
      01-229 1234

      Science Museum
      Exhibition Road, London SW7 2DD
      01-589 4356

Society of Designer Craftsmen
24 Rivington Street, London EC2A 3DU
01-739 3663

Stobart & Sons Ltd
67 Worship Street, London EC2A 2EL
01-247 0501

TRADA   Timber Research and Development Association
Hughenden Valley, High Wycombe, Bucks HP14 4ND
Naphill (024 024) 3091

Triangle Bookshop
36 Bedford Square, London WC1B 3EG
01-631 1381

V&A   Victoria and Albert Museum
Cromwell Road, London SW7 2RL
01-589 6371

Zwemmer's (A.Zwemmer Ltd)
24 Litchfield Street, London WC2H 9NJ
01-836 4710

nb:
Telephone code for central London after May 1990
is 071-

Notespace

22 Advice for beginners

1   Question every brief and *rewrite it* to be sufficiently clear, full, and definite. If you must make certain assumptions of your own, state them. Get agreement for your final version. Then proceed. Fail to do this and you will fail in understanding design.

2   Attitude: if you climb on top of a job, trying to master it, the work will suffocate. *Let it take you, play with it, search for its own life.*

3   Don't be conned into thinking that only new materials or processes are worth investigating. Every material available is strictly contemporary.

4   Out of every job that seems an indistinct mess, try to rescue one small part that is clear, simple, definite, and very well made or done.

5   In many studio projects there is an academic reality – work for you and your tutor – and an 'as-if' reality – work on the job itself. Should you separate them? How?

6   If you are miserably dissatisfied with your work on a job, make your answer to it a detailed self-criticism (a graphic project). Is this an academic reality?

7   Every student understandably begins by striving after originality. After five years work he is delighted if he can attend to a simple job with scruple and insight (unless he is about to launch a successful career as a carrion artist). It helps, at least, to know that.

8   Always show discarded alternatives in support of your work. If you get stuck, develop these alternatives as far as you can.

9   If the world is crowded with inessential rubbish, is there a case for seeing what you can do with the cheapest most simple and most ordinary materials?

10 In the first year remember that you will reject everything before you can complete it, because your values will change so fast. Can you turn this fact to positive account in the way you design? How?

11 Conceive the visible outcome of your work – including notes – as a totality; try to present a running sequential account of your thinking from beginning to end (using diagrams) for someone who doesn't know you or the problem.

12 Use colours freely in a layout pad; if you are beginning, don't sit facing tracts of empty white paper. Study the exact and detailed nature of all given factors in a job, work outwards from them.

13 If you think someone in your group has a better design concept for a job than you have, why not accept and develop it in your own way? The end-result will be very different, and a comparison valuable. You may have the best approach to the next job. Work towards objective standards.

14 In the way of samples or materials or catalogues, collect everything you like or that for some unknown reason holds your attention; not what you ought to like. Information ought to keep pace with your ability to use it.

15 If you must flip through photographs of other people's work, try this: write a short critical commentary on just one photograph, compare notes with someone. You may be surprised at what the eye and intelligence gain from *focus*.

16 Before deciding that Corbusier or Frank Lloyd Wright are thankfully the last of the monumental masons, hitch across Europe or America and see for yourself. On the way, get right off the beaten track and study all the most humble of human artefacts.

23 Questioning design

1   Have you considered the fine distinctions that make human faces recognizable? – the elements are as much alike, and related, as you would get from a problem analysis. So are you just beginning? How do you see a face?

2   Compare gasometers, pylons, cooling towers, street lighting. What is it about cluster high-lighting that makes the rest look sentimental?

3   If you go into a furniture shop, examine the backs and insides and know before leaving how it is (or could be) made.

4   Is a road a shallow kinetic relief? Paved with what intentions?

5   What makes a good street bollard? Geometry, surface, height, spacing, inference, articulation with the ground?

6   On what principles is a street decently furnished? What are the material ingredients?

7   Look at a car and de-gloss it in your mind's eye; what of its form is shine? What about reflection in a glass building? Compare the cut-off of high-rise buildings, vertically and horizontally. Are some visually better than others? If so, why?

8   Is a matchbox fine-tolerance cabinetmaking? Why does it work so well? When should a drawer be a tray?

9   In typography, can you see a Marxist concern for the just allocation of spaces? Or a Freudian concern with motivation and impulse? Where and how in the history of the modern movement?

10 Are milkbottles and wine bottles inferior to art glass from Finland? If so, why?

11 When a textile is used for curtains, what are its functions? What alternatives are there?

12   If you blow up a letter form and cut it out in hardboard, do you like carrying it under your arm? Try setting verse in Plantin, Univers, Bodoni; other things equal, what difference?

13   Should a table be flat? If so, why? Was Albers right to say that flush joints work in metal and not in wood? How would you make a diagram to compactly suggest the nature of wood as a material?

14   Consider noise, signal, redundance; are there distinctions here that would help you design anything?

15   What makes a Georgian sash window 'work' visually and a Victorian sash gape like a fish? How would you replace either?

16   Have you compared drinking cold water from glass, plastic, china, paper, stoneware, metal? Try listing the requirements you would need to satisfy in designing an eggcup and a teapot. Would you need to know much about the flow of liquids? What makes a bad teapot choke or drip?

17   Have you ever seen a well-designed electric fire?

18   One minute exercise: examine that old designer's friend, the Terry anglepoise lamp. Performance aside, what is glaringly inconsistent about it?

19   What is the difference between honesty and sincerity in your work?

20   If you are reorganizing your room, should you 'respect' existing dimensions? If so, why? Are there primary and secondary dimensions?

21   What is the difference between a container and an enclosure? Could it be relevant to know?

22   Would you design a prison? If not, why not?

23   Is a book a four-dimensional object? Should it be seen with X-ray eyes?

24   Can alertness be summoned and sustained by an act of will? If so, when useful?

25 Would you have preferred this book to be illustrated with comic strips? If so, why?

26 Is it good for people to stretch and reach into inconvenient places? If so, is ergonomics a science?

27 What do you like about your favourite things? Compare pubs closely. What makes a good pub? Location, publican, people who use it, decor, privacy, good beer, comfort . . .? Also launderettes. What makes one better to be in than another? Lighting, surfaces, facilities, efficiency . . .? How would you make one better?

28 In a café, consider the tables, etc, drawn out as a plan on your drawing board. Suppose the plan and circulation would read plausibly, the seating comfortable, yet the café is thoroughly depressing to be in. What essential decisions might be missing from your plan? (cf. Le Corbusier, 'the plan is the generator' . . .)

29 What is intrinsically wrong about bookcases, or questionable?

30 Is the surface the heart of things? How else will you know them?

31 Is a door handle a piece of information? What else might you prescribe for an outside door? What is a door, irreducibly?

32 Study watch and clock faces. Should a watch be round, square, or?

33 If you were visually aligning with a circle, would you wish to pick up its centre or periphery? (Other things being equal.) Could they be equal?

34 Is hard architecture male-dominated?

35 Should a well-designed ashtray conceal its contents?

36 Finally, for those who wish to ask why this book has no pictures (as some students have done), here are some reasons. Buried within them are further questions that it may be interesting to frame and ask; questions, moreover, that need answering:

a book conceived as a working tool should not be over-encumbered with precedent;

an effort of imagining, and of search, helps experience to replace consumption;

such a book enters a field to which it is merely contributory, not self-contained;

that field is already image-saturated;

this is reinforced by a tele-media culture tending always to judge by appearances;

the book argues for the primacy of meaning over appearance;

it also laments the influence of magazine and illustrative culture upon the ability of students to design from first principles;

in a book of such generality, an illustrative principle would be falsely finite, closing possibilities rather than opening them;

the book suggests that product design is less adventurous and less definitive than the design of spaces and their equipment;

however, spaces as places defy single-point perspective, needing for proper record a movie camera or supplementary drawings;

publishing cash limits (a real constraint) exclude such possibilities here;

the form of a book should be at one with its message.

Strong counter-arguments can be mounted against each of these points. Do you think they manage to sustain a case?

Most of these questions have invited you to *find out* – by looking, questioning, measuring; then to *sort out* – by comparing, relating, ordering; then to *think out* against such criteria as seem relevant, in context – to make value-judgements.

This is a mouthful.

As would be a description of the designer's world. A world of seemingly disparate and unrelated things, values, interests, persons, processes. Connections have to be found, and made secure. A complex act of reconciliation has to take place; if not of the people and the stones, certainly of the people and the things.

37 Now turn to page 125 (the three stages of diagnosis): a similar statement, but in a context of purpose. Does this help it to make sense?

24 Conference report [1968]

Extract from the Chairman's report on the National Conference on Art & Design Education at the Round House, London NW1, July 1968:

The Conference, which was sponsored by the Movement for Rethinking Art and Design Education (MORADE), had two objectives:

1 to promote whatever action was seen to be called for immediately in view of the present situation in art school throughout the country, and

2 to establish on a national scale the proper bases for further study of art & design education and of the matters that are relevant to it.

Contributors, who were drawn almost equally from the staff and students of art colleges and schools, were invited to speak freely on whatever topics they believed to be important or relevant within the general framework of the following questions:

1 Why art & design education?

2 What is a school of art?, and

3 How should art schools be organized?

The Conference soon found itself to be in agreement that the purpose of art & design education is to develop critical awareness, to allow potentially creative people to develop their aptitudes, to encourage questioning and to stimulate discovery, and to promote creative behaviour. It was also generally agreed that this purpose could not be served except under conditions of freedom far greater than obtain at present – freedom from external control by bodies unsympathetic to and uncomprehending of its purpose, freedom to select students without constraint by irrelevant criteria, freedom to develop courses without regard to inappropriately academic national standards, and freedom from inhibition by too-rigid structures of internal control. The Conference recognized the urgent need for reform by the immediate removal of some impediments but it also recognized that reform in the longer term would need much further study and might well involve the re-orientation of art teaching throughout the educational system as a whole. Voices were not lacking to remind the Conference of the equal need for realism.

A recurrent theme was the relationship of 'Art' to 'Society' and, therefore, of the role or roles – actual and potential – of the artist and designer today. A wide diversity of views was expressed from which it emerged that the need for solidarity in confronting a world unaware of art's value of purposes outweighed the need that might arise for distinguishing differences of function and approach between, say, 'artist' and 'designer'. It was made apparent to the Conference, by the remarks of Sir John Summerson, that even within bodies nominally constituted to represent their views there is an alarming and – in the present situation – possibly crucial lack of fundamental understanding. It was agreed by the Conference, therefore, that a primary function of art education is the extension of understanding and that a world which does not know 'what art is about' will neither be able to use it rightly nor concede to it a proper status. In this 'chicken & egg' situation the need for internal reform is paramount and urgent.

Geoffrey Bocking, Chairman

25   Matchbox maxims [1972]

or   how to keep in business as a troublemaker

:   the following could be a postscript to my book on design education
& procedure

1   In this time formal education is too often a complicated game to
keep us all off the streets; soft at its centre and torpid with false
compliances.

2   Design at its best has an honourable history; affirmative,
questioning, socially and personally committed, seeking to bring
things together in good sense. A very large area of this effort has
degenerated into managerial eyewash.

3   Communication is usually a word to dignify a debasement of
language in the pursuit of profit and the avoidance of plain speaking
in social transactions. Academically, communication theory is
interesting where it is intelligible. A poet might recoil from
communication theory as from a natural enemy.

4   Formal education, design, communication, must now argue for
their life and there is nothing self-evident in the claims of these
activities as now studied or practised.

5   Search: for credibility and a principle (not less a fact) of necessity
in the roots of our work. It is not difficult to be personally creative
whilst culturally inconsequent.

6   Any possible structure for education must be seen, and experienced,
as an accessible hypothesis; argued out and acted out in that way;
tested close to destruction – certainly to continuous renewal.

7   Art and design (as fields of application) are as near together, and
     as different, as priest and doctor would be if we could still respect
     a priestly function. A philosophy – a philosopher – is a man of
     words.

     When we feel threatened, in ourselves or in society, the concerns of
     each may seem identical; and the conditions for health – survival –
     may seem to lie elsewhere.

8   Ask if your environment is strictly necessary – excepting food,
     warmth, shelter, friends. As rhetoric, this is a satisfactory way to
     start. It is also one basis for work assessment.

9   Lethaby: 'design, indeed, design that is really fresh and penetrating,
     co-exists, it seems, only with the simplest conditions'. Consistently,
     he might now add 'recovered and tempered virtue in well-doing'.
     What are our conditions for co-existence?

11   For those who want to investigate in their work, as well as to
     experiment, an intermediate technology is not to be despised.

12   An open hand is a better image than a closed fist.

     From each of these remarks I hope you may find at least one good
     question

This statement was issued to precede a series of lectures to art schools

10   Omitted here for reasons of possible defamation

26　The Bristol experiment

The brief notes that follow concern two phases of the Construction School (1964-79), originally of the West of England College of Art. The work of the School was fully documented, but a detailed account (to be published) would be out of place here.

The School was established by a team of eight people from London (mostly from the Royal College of Art) including a philosopher and an English language specialist, supported, informally, by a group of architects and designers sympathetic to its aims. The work of the group was grafted on to an existing small school of furniture design, with excellent craft traditions, the Head of which (Dennis Darch) now became Convenor of the new school of design. Richard Hollis came as a member of the team to head the then Graphic School, which combined with Construction in a common first year. The group had no career interest in teaching at that time (some members later developed such an interest), but was strongly motivated to explore the following aims: to found a design institute with a teaching component for DipAD, the teaching staff continuing in design practice; to set up a three-year design course without specialization, leading to various areas of design including possibly architecture; and as a school, to set about re-examining certain assumptions or postulates of the modern movement in design. This was a necessary task for design education in 1964 (if not perennially so), following the work of the Hochscule für Gestaltung at Ulm. For certain good reasons (including the skills, equipment, and traditions available) the School concentrated on one-off jobs and special equipment ('problems in which place and occasion were active determinants'[!]) – thus linking to architecture – with the graphic work providing a product-design component in its thinking. Political realities led to the almost immediate abandonment of the institute as an aim; for similar reasons, the intended hospitality to European and American designers had to be curtailed, though interesting work was done with designers such as Paul Schuitema from the Netherlands

The group became interested in a course which, in content, sequence, and outcome, could be irreducibly relevant to designers in all

subsequent fields of specialization; both with respect to knowledge
and technique. Thus a five-year course was postulated, the School
providing the first three, with an optional post-graduate specialization
or the alternative of design office practice: the three years providing
an employable competence. Following the principle usefully and
pithily described by Bruce Archer, that designers should be 'jacks of
all trades but master of one', technical studies were built around
timber practice and technology, for which the teaching skills were
available. However, all projects or specialized areas of study were
intentionally referred back as 'limited cases of general principles',
so that at this time there was no craft bias in the work of the School;
indeed rather the reverse. Accepting students from a foundation
course that was usually fine-art based, the Construction first year
(combined with graphics) was allowed to be intellectually demanding
in contrast, being largely concerned with problem-solving, and with
communication technique. The second year opened out into workshop
practice and technical studies in wood, metals, and plastics. The
third year joined with graphics students again in exhibition design,
but was intentionally student-designed around a series of options
that brought together and developed the work of the first and second
year. In practice, however, the year had to become 'furniture design'
because the School had meanwhile been recognized for this subject.
Despite a most detailed academic programme (which no one could
be found to contest), a highly organized school, enthusiastic staff
and students, and new buildings and equipment, recognition was at
that time confined to specific categories (the relevance of which it was
the purpose of the School to challenge) and in an Alice-like academic
world the School had to apply for recognition in interior design,
which it did not intend to teach, this being the only category with a
five-year expectation. In the event, furniture design was awarded,
thus depriving the School of the student intake that would make the
best sense of its aims, and necessarily altering its subsequent
direction. A national petition, strongly supported by architects
particularly, failed to shift this position. The School therefore set up
a parallel senior vocational course, starting in the workshops and
crossing with the DipAD students half-way. Given excellent students
and a stream of supporting visitors, the School was able to make the
most of its situation, morale was high, and much interesting work
was undertaken. Development at the level, and of the kind, at first
intended, was not of course possible. In 1968/9 (when I left) the
School was able to offer hospitality to Hornsey and Guildford

students, and to contribute support to the international student movement of that time.

For the second phase for which I returned to the School (from 1975 to 1977) – being on this occasion 'voted in' by staff and students – the historical situation, and terms of reference, were very different, and harder to describe in a short space. Various things had gone wrong, morale was low, and the School had lost energy and direction. It was decided to put people and relationships first, in a reorganization that diversified the potential of the School into autonomous groups or communes (intentional communities). Each was to specialize according to the experience and competence of the full-time member of staff in charge; each combined first, second, and third year, in such a way that learning by a form of osmosis could occur in each group; and each built and equipped its own studio environment (sharing technical facilities). The groups were intentionally co-operative within their own family structure, but to some extent competitive within the central area (the arena) where work from all groups was exhibited and criticized, and from which individual student progress was monitored from the standpoint of CNAA requirements. The arena was to be one of critical disputation, but it also had certain co-ordinating and specialized functions (as for tutorials). The Head of the School worked from the arena, with others, but the arena had no power to overrule a genuine autonomy in the groups, which could devise their own working methods (but expect them to be defensible).

Among the objects of this non-heirarchic organization were: the fostering of a student-based and a much more variegated working basis for the School, the mustering and focusing of all available energies in ways that would be at once a challenge and informally derived, the improvement of working and personal relationships all round, the productive working-out of certain internalized tensions that had developed in the School, and a move to make the Head of School rather less indispensable than formerly, his position becoming a critical function and a co-ordinating one. In many ways this experiment proved interesting. Unfortunately a severe staff cut removed the part-time staff who should have circulated between the groups, helping to cross-fertilize ideas, and whose energies were really essential to the success of autonomous grouping; and the School found itself in a vulnerable position for a return visit from the CNAA (having obtained initial support for the experiment in 1976), who did not feel in 1979 that the group system had sufficiently vindicated its rather doubtful promise. At this stage the School was ended (except for those students completing studies).

In retrospect, what did the School achieve? As designers such as Robin Day and Terence Conran were quick to acknowledge, the standard of work was very high in the first phase (though probably too much staff-influenced), and the social structure of the School and its increasingly student-centred learning were the notable contributions of its second phase. In both, the standard of consciousness, of critical discussion, were remarked as being exceptional; and relationships were normally accepted as being adult and responsible even in times of gloom. It could be said, I think with fairness, that when the School was failing it was beyond an average mark-up for success; and this, if true, is greatly to its credit. The employment record for students has been very good and very diverse; an artisan interest has led to several workshops, but without (numerically) a bias in that direction. The School always put architecture first in its design studies, as distinct from its own typical products, and had unusual support from designers in that field; including in the early days Jim Stirling, Peter Burberry, and Richard Rogers, and continuing later with visitors as diverse as Walter Segal, Colin Ward, Ted Cullinan, and John Miller. Colin Boyne (of the *Architects' Journal*) once noted that the first-phase course seemed a very likely foundation for architecture. It is a pity that this bridge-building could not have been taken further (partly for the reasons already given). It has been observed that even emancipated architects tend to distance themselves from 'design' and lean rather heavily on the status-reassurance of their profession; if the work of the School did something to reveal and diminish this problem, this has been a helpful contribution.

Finally, it might be said that an obstinate habit of mind emerged, following Rilke's precept 'hold to the difficult'; in playing it the hard way, the School was always vulnerable, both internally and to the outside world. However, like the AA School of Architecture (with which the Construction School probably had more in common than with most design schools elsewhere), it would be hard for a student to leave without in some useful sense knowing the score. To this extent, at its best, the School suffered the real world. Three prospectus sayings: 'our position is ranged-left and open-ended'; 'putting things together in ways that make sense'; and 'design is a field of concern, response, and enquiry, as often as decision and consequence'.

27   Text references

These references give the source for (where it exists in printed form), or otherwise qualify, material quoted in the text of the book. Place of publication is London, unless otherwise indicated.

page   8   LE CORBUSIER: from 'If I had to teach you architecture', *Focus* (no.1, Summer 1938, p.12). The article is now reprinted in: Dennis Sharp (ed.), *The rationalists* (Architectural Press, 1978).

8   CONRAD: *Victory*; it is Heyst speaking.

8   WOOLLEY: in *Anarchy* (no.97).

11   SCHUMACHER: from *Good Work* (Sphere Books, 1980, pp. 29-30).

25   VITRUVIUS/WOTTON: Sir Henry Wotton, *The elements of architecture* (1624).

25   REICH: epigraph to *The function of the orgasm* (Panther Books, 1968).

33   'freedom': a frequent reference in this text, but not for me to circumscribe. For the distinction from 'liberty', consult Herbert Read's discussion of the matter in *Anarchy and order* (Faber & Faber, 1945), M.L.Berneri's useful *Journey through Utopia* (Routledge & Kegan Paul, 1950), and Erich Fromm's *Fear of freedom* and *The sane society* (Routledge & Kegan Paul, 1942 and 1956).

34   ARCHER: from 'Design as a discipline', *Design studies* (vol.1, no.1, July 1979, p.20).

38   SCHON: see *Beyond the stable state* (Temple Smith, 1971).

38   LUCAS AEROSPACE WORKER: quoted by Mike Cooley, *Architect or bee?* (Slough: Langley Technical Services, 1980, p.80).

39   RICE (quoting Robert Hutchins): in Martin Duberman, *Black Mountain* (Wildwood House, 1974, p.40).

42   LETHABY: in A.R.N.Roberts, *William Richard Lethaby 1857-1931* (LCC Central School of Arts & Crafts, 1957, p.83).

44   WHITEHEAD: from *The aims of education* (Williams & Norgate, 1929, pp. 145, 145-6, 147).

50   GROPIUS: see, for example, *The new architecture and the Bauhaus* (Faber & Faber, 1935, p.92).

51   AUDEN: the lines on 'new styles of architecture' are from 'Petition', first published in his *Poems* (1930) – and not reprinted in all later collections.

51 FRANKL: see *Man's search for meaning* (Hodder & Stoughton, 1963).

53 LARGE: *Sugar in the air* (Jonathan Cape, 1937), seriously due for a reprint – there is nothing of the kind quite so well made.

53 BERGER: for example, *Ways of seeing* (BBC/Penguin, 1972).

56 SCHUITEMA: from 'A statement', *Circuit* (magazine of the West of England College of Art, Autumn 1966, pp. 1, 5). See also his article 'New typographical design in 1930', *New Graphic Design* (no.11, December 1961, pp. 16-19).

75 MEDAWAR: see *Induction and intuition in scientific thought* (Methuen, 1969).

75 POPPER: critics and others who lean upon (their interpretation of) the words of this philosopher, will note the absence of a text reference. In my opinion, Popper is a philosopher's philosopher. I lack the training to estimate the correctness (or otherwise) of his views. Some philosophers (including, oddly enough, Wittgenstein) have the mysterious gift of lay accessibility . . .

85 LETHABY: in Roberts (as above, p.48).

91 NEWMAN: quoted by J.W.N.Sullivan, *Beethoven* (Jonathan Cape, 1927, p.9).

92 WHITEHEAD: from *The aims of education,* (p.18).

102 TURNER: from *Orpheus or the music of the future* (Routledge & Kegan Paul [1926] p.19).

103 ASHBEE on Geddes: from *A Palestine notebook 1918-23,* and quoted by Paddy Kitchen in *A most unsettling person* (Gollancz, 1975, pp. 299, 300).

104 GEDDES on Mackintosh: quoted by Paddy Kitchen in *A most unsettling person,* (p.149).

104 ASHBEE: the earlier statement is from *An endeavour towards the teaching of John Ruskin & William Morris* (Edward Arnold, 1901); both this and the passage from *Should we stop teaching art?* are quoted by Pevsner in *Pioneers of modern design* (Harmondsworth: Penguin Books, 1960, p.26).

104 MUMFORD: cf. his introduction to the 1957 edition of Lethaby's *Form in civilization* (Oxford University Press, 1957, p.xiii).

104 LETHABY: aphorisms, in Roberts (as above, pp. 64, 65, 72).

105 GROPIUS: see his contribution to *Herbert Read: a memorial symposium* edited by Robin Skelton (Methuen, 1970).

106 READ, 'A song for the Spanish anarchists': from his *Collected poems* (Faber & Faber, 1966, pp. 149-50), and quoted with permission of David Higham Associates.

138 'survey before plan': the title of a series of books published by Lund Humphries for the Association for Planning and Regional

Reconstruction; see, for example, *The hub of the house* (no.2, 1945) edited by E.M.Willis.

138 'the plan is the generator': Le Corbusier, *Towards a new architecture* (John Rodker, 1927).

146 RICHARDS: from *Practical criticism* (Routledge & Kegan Paul, 1929, p.182).

207 RILKE: from *Letters to a young poet* (Sidgwick & Jackson, 1945).

Index of names

This index lists names of people and corporate institutions mentioned in the text; those referred to as authors of a book (and in no other context) are not included.

19281 Dillow A14112 —
£10